The AS9100C, AS9110, and AS9120 Handbook

Also available from ASQ Quality Press:

AS9101D Auditing for Process Performance: Combining Conformance and Effectiveness to Meet Customer Satisfaction
Chad Kymal

A Practical Field Guide for AS9100C
Erik Valdemar Myhrberg, Dawn Holly Crabtree, and Rudy "RE" Hacker

The ASQ Auditing Handbook, Fourth Edition
J.P. Russell, editor

Quality Audits for Improved Performance, Third Edition
Dennis R. Arter

ISO 9001:2008 Explained, Third Edition
Charles A. Cianfrani, John E. "Jack" West, and Joseph J. Tsiakals

ISO Lesson Guide 2008: Pocket Guide to ISO 9001-2008, Third Edition
J.P. Russell and Dennis R. Arter

ISO 9001:2008 Internal Audits Made Easy: Tools, Techniques and Step-By-Step Guidelines for Successful Internal Audits, Second Edition
Ann W. Phillips

The Certified Six Sigma Green Belt Handbook
Roderick A. Munro, Matthew J. Maio, Mohamed B. Nawaz, Govindarajan Ramu, and Daniel J. Zrymiak

The Quality Toolbox, Second Edition
Nancy R. Tague

Mapping Work Processes, Second Edition
Bjørn Andersen, Tom Natland Fagerhaug, Bjørnar Henriksen, and Lars E. Onsøyen

Root Cause Analysis: Simplified Tools and Techniques, Second Edition
Bjørn Andersen and Tom Fagerhaug

The Certified Manager of Quality/Organizational Excellence Handbook, Fourth Edition
Russell T. Westcott, editor

To request a complimentary catalog of ASQ Quality Press publications, call 800-248-1946, or visit our website at http://www.asq.org/quality-press.

The AS9100C, AS9110, and AS9120 Handbook

Understanding Aviation, Space, and Defense Best Practices

James Culliton

ASQ Quality Press
Milwaukee, Wisconsin

American Society for Quality, Quality Press, Milwaukee 53203
 Published 2014
Printed in the United States of America
19 18 17 16 15 14   5 4 3 2 1

Library of Congress Cataloging-in-Publication Data
Culliton, James 1946–
The AS9100C, AS9110, and AS9120 handbook : understanding aviation, space, and defense best practices / James Culliton.
pages cm
Includes index.
ISBN 978-0-87389-884-3 (alk. paper)
1. Aerospace engineering—Quality control—Standards. 2. Aerospace industries—Management. 3. ISO 9000 Series Standards. I. Title.
TL671.28.C75 2014
629.102'18—dc23

2014004658

ASQ Mission: The American Society for Quality advances individual, organizational, and community excellence worldwide through learning, quality improvement, and knowledge exchange.

Attention Bookstores, Wholesalers, Schools, and Corporations: ASQ Quality Press books, video, audio, and software are available at quantity discounts with bulk purchases for business, educational, or instructional use. For information, please contact ASQ Quality Press at 800-248-1946, or write to ASQ Quality Press, P.O. Box 3005, Milwaukee, WI 53201-3005.

To place orders or to request a free copy of the ASQ Quality Press Publications Catalog, visit our website at http://www.asq.org/quality-press.

ASQ
Quality Press
600 N. Plankinton Ave.
Milwaukee, WI 53203-2914
Email: books@asq.org
Excellence Through Quality™

For M and all her support.

Contents

List of Figures and Tables

Introduction

This book is designed to give readers an understanding of AS9100C, AS9110, and AS9120. It includes a discussion of each of the standard's clauses, suggests best practices to comply with them, outlines common findings associated with them, and provides an overview of the changes to AS9100C from AS9100B (see Appendix G). This book is in no way presented as a replacement for the text of AS9100C, which is published by SAE International (formerly the Society of Automotive Engineers). The author strongly encourages readers to purchase a copy of the standard from SAE in order to enhance their knowledge and understanding of AS9100C. The standard can be purchased at http://www.sae.org.

The book includes a set of icons designed to help readers determine the type of requirement the section focuses on. Following is a key to these icons:

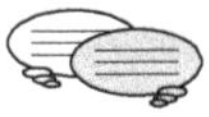 Discussion of a requirement in the standard

 Best practices to follow when implementing the standard

 Findings typically found during audits of the quality management system

Chapter 1
Introduction to Aerospace Quality Management Systems

AS9100, AS9110, and AS9120, the quality management system (QMS) standards for the aerospace industry, are written in the most ambiguous language possible. Indeed, they don't outline how they should be implemented. Those decisions are left to the organization implementing their requirements or, in some cases, to a consultant.

Although some consultant firms for aerospace systems are excellent, there are many that purport to be experts yet proffer systems and processes that are either in contravention to the standards' requirements or so unwieldy that they render the process impotent.

Third-party auditors are not permitted to consult. Still, registration audits can become very heated matters when processes that the prospective registrant has paid a consultant great sums of money to develop and implement fail to meet the registrar's approval.

In an effort to simplify these issues, this book proposes practices that have been described as opportunities for improvement or best practices by registration auditors in the past.

DEFINITIONS AND ABBREVIATIONS

The following terms are used extensively in the aerospace world and are important to understand when operating in the various schemes of certification:

- *Aerospace quality management system (AQMS):* A system implemented to comply with a quality standard issued by SAE International under AS9100, AS9110, or AS9120.
- *AS9100:* A quality management system (QMS) standard designed for the aerospace industry. It was released in October 1999 by SAE and the European Association of Aerospace Industries. AS9100 replaces AS9000 and fully incorporates the QMS requirements of ISO 9001, while adding additional requirements related to quality and safety. Major aerospace manufacturers and suppliers worldwide require compliance with and/or registration to AS9100 as a condition of doing business.
- *AS9110:* Published by SAE, AS9110 is based on AS9100 and adds specific requirements that are critical for the maintenance of commercial, private, and military aircraft. This standard defines the quality

system requirements based on AS9100:2000, with additional criteria for maintenance repair and overhaul facilities that serve the aircraft industry. It was published in January 2003.

- *AS9120:* Also published by SAE and based on AS9100, AS9120 adds specific requirements that are relevant for stockist or pass-through distributors for the aerospace industry. The standard applies to organizations that resell, distribute, and warehouse parts found in aircraft and other aerospace components.
- *AS91xx:* The use of "AS91xx" means that all three AQMS standards apply.
- *AS9104:* This standard applies internationally and defines the requirements for AQMS certification and registration programs. AS9104 defines how the international requirements will be implemented in the Americas Aerospace Quality Group. AS9104 has been restructured into three documents, generally referred to as the "trilogy":
  - AS9104/1: Addresses the basic rules, roles, responsibilities, and requirements of AQMS registration/certification programs.
  - AS9104/2: Addresses the requirements for oversight of AQMS certification/registration programs.
  - AS9104/3: Addresses the requirements for aerospace auditor competency and training courses.
- *ANSI-ASQ National Accreditation Board (ANAB):* The international accreditation body for the United States, which was established in 1989 by the American Society for Quality (ASQ) as the Registrar Accreditation Board (RAB). RAB's original mission was to provide accreditation services for certification bodies (CBs). When RAB was created, it immediately sought to strengthen the US system for CB accreditation by pursuing a formal relationship with the American National Standards Institute (ANSI). In 1991, ANSI and RAB joined forces to establish the American National Accreditation Program for Registrars of Quality Systems. In 1996, with the release of the ISO 14000 family of standards, the ANSI-RAB National Accreditation Program (NAP) was formed, replacing the original joint program.
- *SAE International (formerly the Society of Automotive Engineers):* A global association of more than 121,000 engineers and related technical experts in the automotive, aerospace, and commercial-vehicle industries. Its core competencies are continuing education and standards development.
- *Online Aerospace Supplier Information System (OASIS):* A list of organizations registered to AS9100, AS9110, and AS9120. It includes copies of audit reports and shows the accreditation bodies approved for the various processes, the CBs accredited for the various aerospace schemes, information on all of the auditors approved for the aerospace scheme, and information on registered organizations.

- *Federal Aviation Regulations (FARs):* A set of regulations that control aerospace functions in the United States. It is part of the Federal Aviation Act of 1958 and the Department of Transportation Act of 1960.
- *Federal Aviation Administration (FAA):* The federal agency charged with oversight of all civil aviation, including air traffic control, in the United States.
- *European Aviation Safety Agency (EASA):* Promotes the standards of safety and environmental protection in civil aviation in Europe and worldwide. It is the centerpiece of a new regulatory system that provides for a single European market in the aviation industry.
- *Parts Manufacturer Approval (PMA):* The FAA certificate necessary to allow a manufacturer the right to build and sell aircraft parts.
- *Flight Standards District Office (FSDO):* The FAA field office serving an assigned geographical area and staffed with flight standards personnel who serve the aviation industry and the general public on matters relating to the certification and operation of aircraft.
- *Certificate Holders District Office (CHDO):* The office having direct oversight of any air agency.
- *Manufacturing Inspection District Office (MIDO):* The FAA division with oversight of the manufacturing processes of certificate holders of aircraft, power plant, and constituent parts.
- *Americas Aerospace Quality Group (AAQG):* A cooperative organization within the aerospace industry in North America, Central America, and South America. The purpose of this organization is to establish and maintain a dynamic cooperation based on trust between aerospace companies on initiatives to make significant improvements in quality performance and reductions in cost throughout the value stream.
- *European Aerospace Quality Group (EAQG):* This is the quality community of the Aerospace and Defense Industries Association of Europe and the European sector of the International Aerospace Quality Group. EAQG is a cooperative European global organization of 30 major European companies and 11 national trade associations providing aviation, space, and defense products and services.
- *Asia-Pacific Aerospace Quality Group (APAQG):* The Asian manufacturers that have organized similarly to the EAQG.
- *International Aerospace Quality Group (IAQG):* Its purpose is to implement initiatives that make significant improvements in quality and reductions in cost throughout the value stream by establishing and maintaining dynamic cooperation between international aerospace companies. The AAQG, EAQG, and APAQG are part of the IAQG cooperative. The IAQG is largely responsible for developing AS9100.
- *Industry-Controlled Other Party (ICOP):* The aviation, space, and defense ICOP scheme was developed as a means for industry stakeholders to

rely on QMS certificates as a component of their supplier approval and surveillance processes. It includes oversight activities of registration bodies and auditors.

- *Instructions for continued airworthiness (ICAW):* Maintenance instructions approved or accepted by the FAA for aeronautical products.
- *Transportation Security Administration (TSA):* Creates rules for FAA-approved repair stations operating under FAR 145.
- *Minor finding:* A single lapse or breakdown in a process that is not likely to result in the failure of the quality system or reduce its ability to ensure controlled processes or products.
- *Major finding:* The absence or total breakdown of a system to meet a requirement. A number of minor nonconformities against one requirement can represent a total breakdown of the system and thus be considered a major nonconformity or a condition that may result in the failure, or materially reduce the usability, of the products or services for their intended purpose. A noncompliance that judgment and experience indicate is likely either to result in the failure of the quality system or to materially reduce its ability to ensure controlled processes or products.
- *Special requirements:* Requirements identified by the customer or determined by the organization that have higher risks of being achieved, thus requiring their inclusion in the risk management process. Factors used in the determination of special requirements include product or process complexity, past experience, and product to process maturities. Examples of special requirements include performance requirements imposed by the customer that are at the limit of the industry's capability and requirements determined by the organization to be at the limit of its technical or process capabilities.
- *Critical items:* Those items that have a significant effect on product realization and use of the product (e.g., functions, parts, software, characteristics, processes). This includes safety, performance, form, fit, function, producibility, and service life, which require specific actions to ensure that they are adequately managed. Examples of critical items include safety-critical items, fracture-critical items, mission-critical items, and key characteristics.
- *Key characteristic:* An attribute or feature whose variation has a significant effect on product fit, form, function, performance, or producibility; specific actions are required to control variation.
- *Original equipment manufacturer (OEM):* One of the prime manufacturers that rely on the aerospace supply chain for manufacture, repair, and service of aircraft and constituent parts.

Aerospace-Specific Terms and Acronyms

- *Alternate means of compliance (AMOC):* An FAA-approved method of compliance with an airworthiness directive in a manner other than stated within the airworthiness directive.
- *Aviation maintenance organization (AMO):* The Canadian equivalent of an FAA repair station.
- *Type certificate (TC):* The FAA approval necessary for each type of aircraft, engine, propeller, or rotor. The TC is the basis for approval for OEMs to produce the aircraft under the terms of a production certificate.
- *Production certificate (PC):* A certificate issued by the manufacturing inspection district office for the production of the aircraft after it has been approved by the aircraft certification office and given a TC.
- *Supplemental type certificate (STC):* An approval for a modification to an existing TC. It may be held by the OEM or a third party.
- *Air carrier:* A person or organization that undertakes directly, by lease or other arrangement, to engage in air transportation.
- *Designated engineering representative (DER):* An individual approved by the Aircraft Certification Office to review and recommend approvals or give direct approvals for design, modification, and repair.
- *Designated airworthiness representative (DAR):* An individual approved by the Manufacturing Inspection District Office and/or flight standards to approve installations, modifications, and repairs.
- *Organizational designated authorization (ODA):* An individual at a repair station who may approve data for major repairs and alterations, issue airworthiness certificates and approvals, and perform aging aircraft inspections and records review.
- *Aircraft Certification Office (ACO):* The part of the FAA responsible for design approvals.
- *Equivalent level of safety (ELOS):* A term used to ensure that modifications or changes to a design and processes have the same level of safety as the original terms of certification.

FINDINGS

The process of AS91xx certification and this book use the word "finding" prolifically. It is the intent of registrars and most auditors to conduct audits with the objective of the organization showing conformity to the respective standard. When a process is shown to vary from the requirements of the standard, a finding of such variation is written against the certificate. The findings are classified as minor or major (see definitions of major and minor findings given earlier), and the organization is expected to respond to the finding with a thorough root cause analysis and propose an action plan to correct the (variation) finding. The time frame for response is dependent on the registrar's documented procedures and/or the specific time constraints imposed by an expiring certificate. This book lists the common findings shown to the author over many years of auditing aerospace organizations.

COMMONLY CONFUSED PROCESSES

The transition from ISO 9001 to any aerospace standard is easier than acquiring a registration from no prior QMS; however, there are significant changes in the processes that require much thought, drafting, and implementation to be effective. New registrants tend to overwrite procedures and processes. This tendency extends to some consultants who showcase a larger, more complicated manual system—versus a slimmer, more efficient system that is tailored to the operation—as an incentive to charge a higher price.

There is no cookie-cutter product available that will meet all aerospace firms' needs in attaining and operating under one of the three aerospace schemes. An organization seeking registration should tailor the requirements of the standard to its operation in the most efficient manner possible. There are small shops that have burdensome corrective action processes that would tax even a large organization for compliance. The result of this mismatch of organization to procedure is a finding. In some cases these are repeat findings, which, in turn, result in major findings.

Major findings are not only embarrassing to the quality department, the organization, and top management; they are also frightfully expensive. They result in follow-up visits that are in excess of the planned registrar fees. A two-day follow-up visit with transportation, expenses, and fees can cost $5000 or more. To make matters worse, this appears on the OASIS database for all OEMs to see.

It's important, then, to maximize the efficiency of your QMS and reduce the likelihood of findings so that the system will be productive and rewarding to the organization.

The AS9101D Enigma

When AS9100C was rolled out, it included a plethora of new processes, procedures, and requirements that were not directly related or apparent to the organizations using or becoming certified to AS91xx.

Among the impacts was the elevation of the ICOP process to a wider and more far-reaching responsibility in the certification and oversight of the AS91xx schemes. Problems in OEM supply chain management have resulted in the concept that, in addition to increasing the requirements of AS9100C, poor supplier performance must be the result of poor registrar and/or auditor performance, ability, and training.

OEM oversight appeared to show a trend of high numbers of OEM findings being made on suppliers that either failed to deliver on time or had high numbers of defective products, whereas there were no corresponding numbers of findings in those same organizations by the registrar (CB). In fact, some of those same suppliers that received numerous OEM findings were not receiving any findings by the registrar.

Obviously there was a breakdown somewhere. While most OEM auditors are "product" auditors and the CB auditor is a "systems" auditor, there is a correlation between a breakdown in product and a defective system.

The result of this oversight discovery caused several things to happen collaterally with the introduction of AS9100C:

1. The trilogy of AS9104 was completed with the introduction of AS9104/1 through two supplemental rules from AS9104A.
2. Training was mandated for all aerospace auditors before they could perform audits to AS9100C.
3. Training was allowed only through an approved "trainer" (e.g., Plexus).
4. The auditor had to pass the course with a score of 80% or higher.
5. An auditor who failed to make 80% could retake the test up to two more times. If the auditor was still unable to pass, he or she had to retake the training. (The training isn't cheap. The course runs about $2500 plus expenses, and the retake is $750.)
6. The old AS9101C checklist, which comprised the report for AS9100B, was reintroduced as AS9101D, a stand-alone aerospace standard.
7. AS9101D contains 24 pages of instructions and 7 checklists, some of which are used more than once on any given audit.
8. The OASIS database was ramped up not only to capture portions of the CB reporting but also to include the complete report, track the performance of the auditors, and ensure compliance with certification deadlines.
9. The ANAB was ramped up to require more witness audits of auditors in the field.
10. ICOP oversight of CBs was ramped up to ensure report content and a commitment to comply with the new requirements.
11. Reporting of the audits changed from element based (on a clause of the standard) to process based (on the basis of the performance of the process).
12. Audit duration requirements were changed, and the addition of the audit structure was implemented (see "Certification Structure" section).

All these changes were made without much communication to the organizations themselves, so they all went ahead thinking that AS9100C was just a simple revision of the standard, much like AS9100 to AS9100B.

The common response from the organization was: "I am being certified to AS9100C, not AS9101D. I have never heard of AS9101D and don't want it."

And so it began . . .

Now that the revised system has been in place, most organizations understand the relationship that AS9100C is their standard and that it is administered by a qualified CB using properly trained auditors with the appropriate oversight and using AS9101D as the certification rulebook.

Certification Structure

Before AS9100C, only two types of certifications were allowed: single sites and, if conditions were met, multiple sites.

With the implementation of AS9104/1, the rules were clarified and there are now five types of structures:

1. Single sites
2. Multiple sites
3. Campus sites
4. Several sites
5. Complex sites

The requirements for each type of structure are far beyond the purpose of this handbook, but they may be found in Appendix B of AS9104/1. The structure must be agreed on by the CB and the organization and, in the case of a complex site, approved by the Other Party Management Team prior to use.

THE AS91XX AQMS

Since replacing AS9000 in 1999, AS9100 revision B and now revision C have demonstrated their effectiveness through more consistent verification methods and expectations as well as fewer verification audits for the industry from customers, OEMs, and industry associations such as the Coordinating Agency for Supplier Evaluation (CASE). The result has been improved supplier performance and reduced oversight costs. In 2003, AS9110 was published, and repair stations were given the opportunity to rid themselves of the manufacturing concepts of AS9100 and be certified to a scheme that meets their particular needs.

OEMs within the aerospace industry use a variety of verification requirements. Some use internal auditors to verify suppliers' QMS, and others share their quality system audit results with industry suppliers. More recently, OEMs have incorporated the ICOP scheme as an additional tool for supplier oversight through increased registrar oversight. ICOP has an online presence in the OASIS database. This oversight program has resulted in rule changes and additional requirements imposed on auditors and registrars.

It's important to note ICOP's concern about the practice of "soft scoring" aerospace audits. This practice is a holdover from the AS9101C checklist as well as AS9110 and AS9120 and their checklists (AS9111 and AS9121, respectively); it is based on a 1000-point scale for organizations that have full registration to the standards. Organizations that exclude design and development from their scope are scored on an 880-point scale. During audits, findings result in deductions from the points on the scorecard.

Soft scoring occurs when auditors don't list all of the nonconformities found during an audit, which results in an artificially high audit score. Because it misrepresents the true score attained by an organization, the organization receives an unfair advantage over properly scored organizations. Since 2008, there has been a push back on this practice, resulting in generally more findings. This makes it more important for organizations to implement straightforward procedures that comply with the requirements of standards.

This book does not detail all of the requirements in the standards that apply to aerospace companies. Many of the processes they are required to implement are fairly simple and straightforward. In some cases, simple notations are the best practice available. Furthermore, there is no one correct way to comply with any AQMS requirement. Auditors who assert that there is only one way to comply with a requirement of any standard should be challenged (politely, of course). If satisfaction cannot be attained at the auditor level, the registrar should be contacted. Appeals processes are mandated by the ANAB. It's important that your processes serve your purpose and also satisfy the scheme requirement. Never write a procedure or implement a process merely to satisfy a standard's requirement. Always ask yourself, "What value will this procedure add to my organization?" in order to avoid drafting a convoluted procedure that requires multiple approvals, summary reports, required attendance, or record keeping. To wit, AS9100 states:

> The adoption of a quality management system should be a strategic decision of an organization. The design and implementation of an organization's quality management system is influenced by varying needs, particular objectives, the products provided, the processes employed, and the size and structure of the organization. It is not the intent of this international standard to imply uniformity in the structure of quality management systems or uniformity of documentation.

This is why such care must be taken in designing and implementing your AQMS. The overall scheme has one overriding tenet: the plan-do-check-act (PDCA) cycle (Figure 1.1). This can be briefly described as follows:

- *Plan:* Establish the objectives and processes necessary to deliver results in accordance with customer requirements and the organization's policies
- *Do:* Implement the processes
- *Check:* Monitor and measure processes and products against policies, objectives, and requirements for the product and report the results
- *Act:* Take action to continually improve process performance

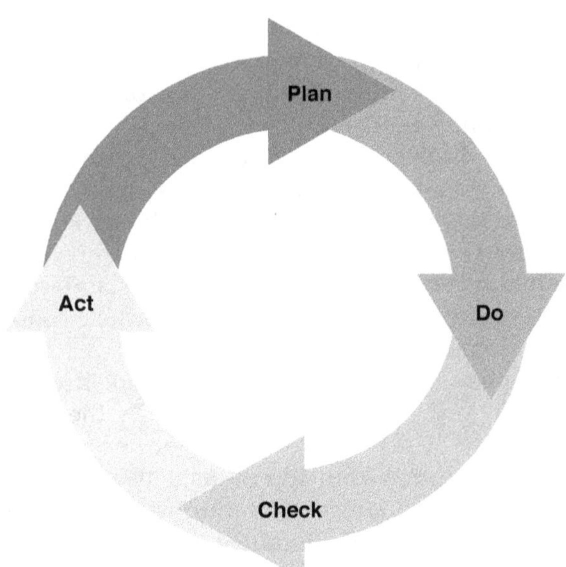

Figure 1.1 Plan-do-check-act cycle.

This book focuses on the PDCA's "plan" function to help organizations develop and implement procedures and processes that fit them. It also discusses common traps and pitfalls in the AS9100C requirements that are likely to result in findings.

THE STANDARD'S REQUIREMENTS

AS9100C's requirements begin with section 4.0. Sections 0, 1, 2, and 3 are explanations of and introductions to these requirements. The important concepts in the introductory portions of the standard include new definitions that will help readers understand its new requirements. Revision C's most important new requirement relates to risk management, which is addressed in section 7.1.2. This may initially seem to be an onerous requirement, but close inspection of the section and its definitions of "collateral concepts" make it easy to understand.

Sections 4, 5, and 6 of AS9100C contain few major changes. Smaller aerospace firms and those that are new to third-party registration might find some of the sections' requirements unfamiliar. For example, section 4 requires the organization's quality manual to include a list of interactions of processes. This is an area that is commonly missed or misunderstood. Section 4.2.2 states:

> The organization shall establish and maintain a quality manual that includes:
>
> a. The scope of the quality management system, including details of and justification for any exclusions (see 1.2)
>
> b. The documented procedures established for the quality management system, or reference to them
>
> c. A description of the interaction between the processes of the quality management system

This requirement has been interpreted to mean both how an organization processes a customer's order and which business functions are available to customers. Figure 1.2 illustrates the standard's requirements for process interaction.

This is a simple explanation of the interaction of processes, but it's not the only interaction description available. Custom process maps and flowcharts, some of which are better than the standard's depiction, can also be used. In any case, make the depiction of process interactions as simple as possible, but note that section 4.1 warns users to take this requirement seriously. It is a key ingredient of certification and a major part of the audit process. It's usually easier to become certified than it is to maintain certification because of the standard's continual improvement constraints—especially under revision C, because continual improvement is determined to be a stand-alone process.

All three aerospace standards and ISO 9001 make frequent references to "top management." Many organizations simply identify their CEO or president with this title, but it's wise to make sure that he or she is the right person for the required job. Make sure that your top management is truly responsible for the various sections of the standard that require top management's oversight. Top management may be a steering committee, an organizational level, department heads, division heads, or a series of individuals.

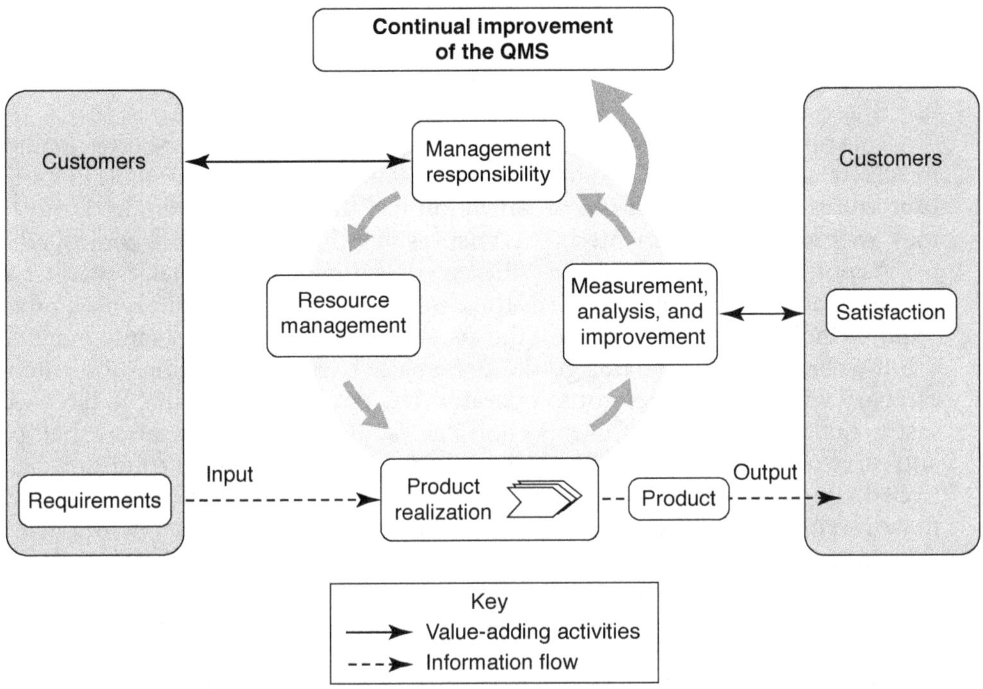

Figure 1.2 Interactions of process-based quality management.

AS9101D requires that top management be interviewed at each audit. Accordingly, ensure that top management personnel are familiar with their responsibilities under AS9100C.

REQUIREMENTS AND BEST PRACTICES

The fact that the standard specifically mentions only six documented procedures often leaves the impression that it doesn't require any other documentation, procedures, or written material. This is not true. Note section 4.2.1(d):

> The quality management system documentation shall include:
>
> a. Documented statements of a quality policy and quality objectives
>
> b. A quality manual
>
> c. Documented procedures required by this International Standard
>
> d. Documents, including records, determined by the organization to be necessary to ensure the effective planning, operation, and control of its processes

This means that the requirements that call for "criteria" (section 7.4.1), "establishment" (section 5.4.1), or "definition" (section 7.6) must be documented in the form of processes, procedures, or work instructions even though they are not required as "documented procedures."

A key document in the overall audit scheme is ISO 19011. This standard is the guidance document for auditors in the performance of audits. Section 3.3 of this standard defines audit evidence as records, statements of fact, or other information that is relevant to the audit criteria and *verifiable*.

This infers that processes other than the six documented procedures included in AS9100 should be in place to verify compliance. Auditors will look to various procedures to ensure that the organization has the processes required and that they are properly implemented. This ensures that the processes are controlled.

A controlled process is one in which work is done in a designated manner and sequence and consistently achieves the desired outcome. Uniformity in a process requires that each worker perform the tasks assigned in the exact same manner.

If workers don't have this guidance (usually a work instruction), the outcome can vary widely from operator to operator. This is why it's essential to have work instructions at the point of use. Section 7.5.1 states: "The organization shall plan and carry out production and service provision under controlled conditions. Controlled conditions shall include, as applicable, a) the availability of information that describes the characteristics of the product, b) the availability of work instructions, as necessary."

This seems to be a troubling requirement for many registrants who have been told only of the six required documented procedures. It's an important misconception, but one that is easily fixed.

This is the principal goal of this book: We will discuss all three AQMS standards (along with a short discussion of their requirements), pose potential questions auditors may ask, explore common findings that are generated in the clause area, and suggest best practices. Remember that any process or procedure should be tailored to your specific quality system.

NOTES

Chapter 2
Risk Management: Sections 0–3

The requirements in AS91xx are complementary. Thus, any regulatory, contractual, and/or customer requirement will take precedence over any requirement of AS9100C. This is an important concept and not to be taken lightly. You will likely have to explain to your auditor why contractual and customer requirements override an AS91xx requirement.

Section 1.2 of AS9100C includes guidance on when AS9100, AS9110, or AS9120 should be the selected standard used by an organization. The International Aerospace Quality Group (IAQG) believes that quality management systems (QMSs) can benefit from the implementation of AS9100C by non-aerospace systems and could possibly receive additional recognition and synergies with NATO allied quality assurance publications.

Section 3 contains new definitions that are important to understanding AS91xx. These definitions should not be thought of as stand-alone concepts. Each of them has its place in the overall flow of the requirements in AS91xx. Much has been made of the concept of risk management contained in section 7.1.2. Many organizations are somewhat concerned about what the registrar will require for compliance with this section, and there is very little information available about it.

The important definitions include the following:

- 3.1 Risk
- 3.2 Special requirements
- 3.3 Critical items
- 3.4 Key characteristics

The concept that risk management is part of the relationship between special requirements, critical items, key characteristics, and configuration management is an important part of AS9100C.

Because critical items are often part or a subset of a deliverable product, one of the first items of planning should be the identification of specific features or characteristics that are most important to the customer. These characteristics could be critical, major, or safety oriented, but in AS9100C they are simply called "key characteristics."

Special requirements and critical items must be considered at all phases of the product life cycle: through review of requirements, design and development inputs, design and development outputs, purchasing, control of production, and, finally, monitoring and measurement.

The relationship of these terms, as well as the concept of risk management, is embodied in the explanation of each of the following new definitions:

- *Special requirements:* Requirements identified by the customer or organization that have higher risk and significant importance to product performance, safety, delivery, or customer satisfaction. Special requirements translate to critical items. Examples of special requirements include the following:

  —Special processes and performance requirements imposed by the customer

  —Requirements that are new or unknown to the organization

  —Requirements that may exceed capacity

  —New or future regulations or requirements determined by the organization to be at the limit of its technical or process capabilities

  —Any other requirement that is essential to program success

  Special requirements can originate from customers, regulatory authorities, environmental conditions, political conditions, economic conditions, and/or market conditions. They can also flow from internal conditions, such as the organization's capacity, workforce, facility, available equipment, and financial condition. Customer-specified special requirements are most often identified during the contract review. Internally defined special requirements are most often identified during the design process. Special requirements related to process capability and capacity might be determined during the realization planning stages of section 7.1.

- *Critical items*: Items that have a significant effect on product realization. These are most often parts but can be processes and software. In the use of the product (including safety, performance, form, fit, function, productivity, and service life), they require specific actions to ensure that they are adequately managed. Examples of critical items include the following:

  —Safety-critical items

  —Fracture-critical items as determined by customer and design requirements

  —Mission-critical items

  —Key characteristics

These lists need to be well understood and defined during product realization, as they interrelate with the concept of risk management. Risk management isn't a separate program function of the overall program planning and management process. To be effective, the risk management process must be recognized as a program management activity and not something limited to the engineering function. Any program element associated with cost, schedule, and performance has a direct interface with a risk management process.

It's important to remember that risk management is employed throughout the program's life cycle. The risk management strategy should be developed early in the program, perhaps as early as stage zero, and addressed continually. This process doesn't change fundamentally as the program progresses, although it should be refined as the program's "unknowns" become "knowns." Figure 2.1 shows the relationship of these elements to the overall risk concept.

Special requirements should be an input of risk management, but it's important to note that there isn't just one method for controlling or managing special requirement risks. The type of risk and potential consequences will likely require different treatments, mitigation, and/or contingency plans. The processes to accomplish this should be documented as well as the outcomes of such processes.

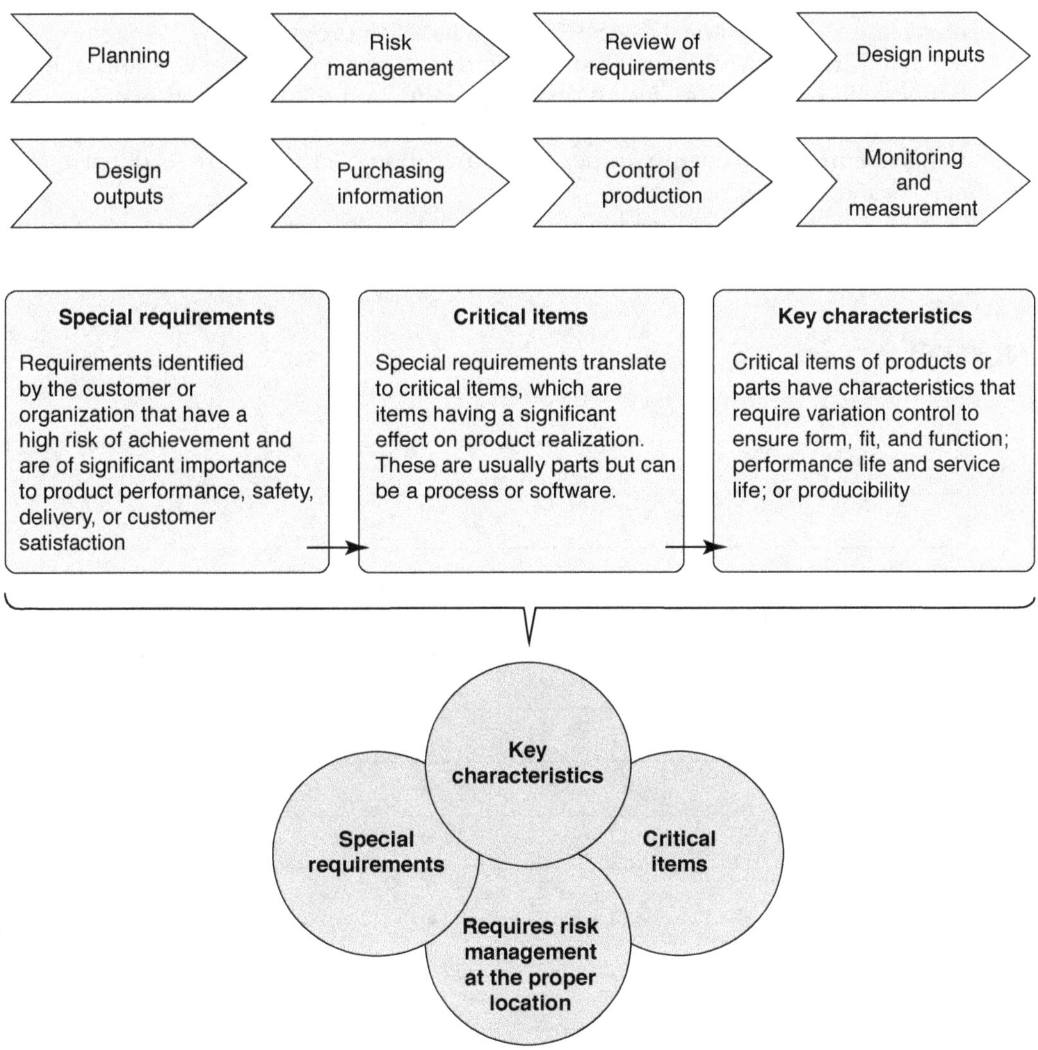

Figure 2.1 Process phases.

BEST PRACTICES

Manufacturing capability issues may require an organization to invest in new facilities, equipment, training, or personnel. These issues might include clean rooms, new equipment, or additional process controls. Physical threats, such as environmental issues, may require development of a business recovery plan. Knowledge and skills concentration within a small organization may require development of a formal training plan for others. Customer-specific requirements, such as the imposition of Nadcap, may require development of formalized plans to accomplish the goal.

Critical items may be explicitly identified and marked by the customer in the request for quote, request for proposal, statement of work, contract drawings, and/or specifications. The organization should have a proper understanding of the identified critical items in concurrence with its customer. Customer requirements are an integral part of AS9100C and its audit program.

Even if the customer has not explicitly defined critical items, the organization should evaluate the design outputs, production planning, and operation for potentially critical items. Critical items should be explicitly listed in the organization's documents to ensure proper communication of criticality to its departments and/or individuals.

It's a good practice to identify critical items on work orders or travelers to avoid a slowdown of this requirement through all phases of product realization.

NOTES

Chapter 3

Aerospace Quality Management Systems and Documentation

SECTION 4.1: GENERAL REQUIREMENTS

Discussion

This section requires that the organization obtain and maintain certification from the responsible authority, whether it is the Federal Aviation Administration (FAA), Transport Canada, or other national approval. It imposes the inclusion of AS9100's requirements into your quality management system (QMS). One keystone requirement is paragraph (d), which requires the availability of resources to support the operation. Too many times auditors are told that the reason for findings is the "lack of resources." This could lead to additional findings if sufficient resources are not available. Never allege that the root cause is a result of inadequate resources. Section 4.1 also imposes the amount of control that you have over outsourced processes. These include tool calibration, plating, nondestructive testing, painting (which is sometimes a special process under subclause 7.5.2), welding, and assembly.

The requirement for determination of the sequence and interaction of the processes, criteria, and methods needed to ensure that the operation and control of processes are effective may take specialized assistance in the form of subject matter experts and tools such as process maps or flowcharts.

In any case, this sets the stage for all the processes you will use in your AS91xx operation. This is typically a management function but may include employee assistance in various stages of the planning. Note that the first paragraph of this section requires documentation. This is a recurring theme. The standard requires you to implement actions necessary to achieve planned results (e.g., quality objectives, product conformity, and customer satisfaction) and continual improvement of these processes.

Any organization that outsources processes that affect product conformity to a customer, regulatory, or statutory requirement must ensure control over such processes. The type and extent of control applied must be defined within the organization's QMS. Questions frequently arise over what constitutes an outsourced process. Some examples of frequently missed outsourcing include tool calibration, records retention, packaging, and shipping.

Findings

As stated earlier, the most common finding in section 4.1 comes from the lack of resources. Consistent customer complaints about on-time delivery or quality, failure to meet quality objectives of the same two elements and repeated escapes, nonconforming product, and customer rejections can lead to major and minor findings.

Best Practices

You must be able to show auditors that there is a process for planning processes, such as traveler production, resource planning, and continual improvement. A process map is the best tool available to demonstrate this requirement. To this end, be sure you have a good grasp of what a process map is. Process maps should be simple but descriptive enough to show the sequences and interactions. Is there a process map that shows how the entire process links together and how the overall process links corporate, sales, design, manufacturing, assembly, and the warehouse? Does your process map show the sequence and interaction of the processes at your facility? Do the inputs and outputs match? Does the process interaction make sense relative to the process being studied?

Auditors often believe that top management views the implementation and auditing of AS9100 (and ISO 9001) as necessary evils that are dictated by original equipment manufacturers (OEMs). Thus, they think that management overlooks the standard's value, relevance, and importance. However, one value-added area emphasized in both AS9100 and ISO 9001 is the process focus. Implementing and auditing a process-focused QMS will provide a tremendous value to the organization and reduce the amount of work. The process approach is also a fundamental building block of ISO 9001:2008 and AS9100, AS9110, and AS9120, as noted in the following passage from ISO 9001:2008:

> This International Standard promotes the adoption of a process approach when developing, implementing, and improving the effectiveness of a quality management system to enhance customer satisfaction by meeting customer requirements.
>
> For an organization to function effectively, it has to identify and manage numerous linked activities. An activity using resources and managed to enable the transformation of inputs into outputs can be considered as a process. Often the output from one process directly forms the input to the next.
>
> The application of a system of processes within an organization, together with the identification and interactions of these processes and their management, can be referred to as the "process approach."
>
> An advantage of the process approach is the ongoing control that it provides over the linkage between the individual processes within the system of processes, as well as over their combination and interaction.

You can see how all this will fit together. Begin your journey to AS91xx certification with the process approach and apply it to all processes—especially internal auditing; you will find the process to be considerably easier to obtain and maintain.

Make sure you have management and employee buy-in on these concepts and encourage these individuals to participate.

SECTION 4.2.1: DOCUMENTATION REQUIREMENTS—GENERAL

Discussion

Auditors and their auditee organizations sometimes disagree about the absence of documentation and procedures when the auditee presents only a quality manual and the six documented procedures specifically required by AS9100. These differences usually arise from differing interpretations of the requirements contained in clause 4.2.1 and its related notes, which require the QMS documentation to include documents needed by the organization to ensure effective planning, operation, and control of its processes. The extent of the QMS documentation can differ from one organization to another due to the following:

- The size of the organization and the type of activities
- The complexity of processes and their interactions
- The competence of personnel

There should never be a conflict between the auditor and the organization, because the object of certification to an aerospace quality management system (AQMS) scheme is a partnership approach between the organization and its registrar. The key question here concerns the organization's operating processes. The need for documentation is the need for consistency. Documentation helps organizations identify and avoid significant risks. An analysis of your operation should yield procedures, forms, records, and work instructions to provide the consistency needed to satisfy the customer, particularly if the customer is an OEM and the Industry-Controlled Other Party (ICOP) scheme applies.

Findings

Most findings in this clause result from either inadequate or conflicting process documentation. Other than omission of the required documentation—such as a quality policy, a quality manual, and/or the six documented procedures or records (which has been known to occur)—the predominant findings generally concern work instructions, logs, or records that are required by the procedures.

Best Practices

To fulfill the requirement that logs be fully evaluated, ask whether a log is a value-added necessity that outweighs the time required to complete, maintain, and archive it.

There is no longer a requirement for a separate document that references the relationship between the requirements of the standard and the organization's documented procedures. This is not to say such a document won't be helpful in finding the relevant procedures if they are numbered differently than the requirements in the standard.

This section requires documentation in excess of the six documented procedures already explained. You must create documentation that includes the procedures necessary to ensure the effective planning, operation, and control of the processes. Ensure that your documentation is appropriate to the size of your organization.

A common question about this section is, "Do I need a written procedure for contract review?" The answer is that it's not one of the six required documented procedures in the standard. However, it will be difficult for the customer service department or a contract reviewer to process contracts without written guidance to ensure that whoever performs the function follows the correct procedures. This means some kind of written process or procedure.

SECTION 4.2.2: QUALITY MANUAL

Discussion

The "scope of the QMS" is a term commonly used to describe the organization's processes and products (and/or services), related sites, departments, divisions, and so on, to which it applies a formal QMS. The scope of the QMS should be based on the nature of the organization's products and its realization processes; the result of risk assessments and commercial considerations; and contractual, statutory, and regulatory requirements. This doesn't necessarily include all the processes, products, sites, departments, or divisions of the organization.

Many organizations seem to read past this requirement, so it's a common source of unnecessary findings. This section simply requires a quality manual, the six documented procedures (or reference to them), and an interaction of processes. The requirements for the scope and justification of exclusions should be straightforward.

It's a tremendous waste of time, effort, and resources to create a quality manual that merely regurgitates the requirements of a standard. Section 4.2.2 outlines AS9100C's requirements for a quality manual. If you follow this section to the letter, your entire quality manual will be shorter than four pages. Its only requirements are the scope of the QMS (including details of and justification for any exclusions), the documented procedures established for the QMS or reference to them, and a description of the interaction between the processes of the QMS.

Findings

Findings in this area are usually administrative. A statement that the exclusion taken for design and development is "N/A" is grounds for a finding. The requirement is to describe the "details and justification" of exclusions. Why is it excluded? Likewise, the absence of a process map or interaction description is also grounds for a finding.

Beware of conflicting information. Some organizations will schedule annual management reviews and quality manual reviews, but in their procedure state that these should be done biannually. Ensure that planned and required schedules align.

Best Practices

Make sure that the elements required by section 4.2.2 are listed in the quality manual. Make clear statements as to the justification for exclusions, for example, "Design and development is excluded because this organization possesses no design service and is a production facility that builds to the customer's specifications" or "Servicing is excluded because this organization does not service components after manufacture and shipping to the customer."

Draft a good scope statement. Don't offer a statement such as "builds aircraft parts." Try a simple expansion, "Boogaloo Machine Inc. is a precision machine shop specializing in prototype, tooling, short run, and production machine work for the high-tech, aerospace, telecom, and other industries."

The organization must decide whether to incorporate the quality manual into the FAA repair station manual/quality system or write a separate and stand-alone document for AS9110. If possible, the quality manual should not be incorporated into the FAA repair station manual. The main reason for this is that under FAA rules, every change made to the quality manual must be submitted to the FAA for acceptance. The organization can avoid this if the AQMS manual is a stand-alone document.

SECTION 4.2.3: CONTROL OF DOCUMENTS

Discussion

Control of documents is the first of the six required documented procedures and probably the greatest source of findings in any AQMS certificate audit program. Document control is a cornerstone of any AQMS. The requirement seems simple: "Documents required by the quality management system shall be controlled." Still, most business operation errors are the result of paperwork errors.

New and prospective AS91xx registrants are most often confused by the standards' requirement for documented procedures. The need to control external documents is particularly troublesome. External documents may be customer documents or part of an external standard, such as Mil-Specs or other industry standards. In the case of AS9110 it can be a component maintenance manual (CMM), an overhaul manual (OHM), or aircraft, engine, or propeller/rotor maintenance documentation.

Although helpful, electronic document systems have problems of their own. Given the increased capacity to modify, update, reformat, and otherwise alter documents within an electronic-based management system, it's important to address attention to the control elements (e.g., document identification and document revision level). As electronic media have led to an increased rate of document modifications, ensure that obsolete documents and modification permissions are carefully managed.

Findings

The most common finding in this clause regards revision-level control of documents. For example, work instructions in use are found to be outdated by three revision levels. Major findings are written because of a large number of minor findings in this section, as it constitutes the breakdown of a required process.

The procedure itself is a source of findings for missing elements, such as external documents or electronic documents. Backup of documents (and records) is a requirement where those processes are utilized. Disaster recovery plans should be in place to replace the document control system in the event of catastrophe.

There are eight levels in the life cycle of AS91xx quality documents:

1. Creation
2. Approval
3. Distribution
4. Review and update as necessary
5. Change
6. Re-approval
7. Redistribution
8. Obsolescence

Adherence to these steps should save many findings during systems audits.

Best Practices

Many organizations employ electronic systems that lock out obsolete documents. If your system doesn't, you must ensure that any obsolete documents are removed from work areas. Technicians have a habit of posting a drawing or specification for repetitive operations at the workstation, preventing effective document control. Always look for uncontrolled documents and review the master document or form list prior to audits to ensure proper revision levels. Remember that document control is an integral part of configuration management.

Make sure that all document users are aware of the control requirements. Major findings have been issued for repair/overhaul or testing final products to the wrong revision level and releasing them to the customer.

SECTION 4.2.4: CONTROL OF RECORDS

Discussion

The requirements for the documented procedure must include your policy to define the controls needed for the identification, storage, protection, retrieval, retention time, and disposition of records. There is no direct guidance on retention times. The FAA requires that records be kept for two years. This can be an important benchmark for your decision under AS9110. The requirement for controlling supplier records is often missed, a problem that's fairly unique to aerospace companies. Most companies doing business in this industry are longtime experts on document retention and records, so times for production records and materials are usually dictated in contracts between customers and suppliers. If your records are kept off-site, the audit will probably be longer to accommodate their review at the storage location. Finally, if you outsource the storage of records, you must disclose this in the quality manual and the facility must be treated as a vendor.

Records are the best evidence of a functional system. They demonstrate the effective operation of a QMS. If there is no record, it never happened.

AS9120 has specific record-keeping requirements:

> Records of product origin, conformity and shipment shall be maintained in accordance with customer, statutory and regulatory requirements.
>
> NOTE: Records include but are not limited to manufacturer, distributor, repair station, test and inspection reports; certificates of conformity (manufacturer, sub-tier distributor), copies of airworthiness certificates; nonconformance, concession and corrective action records; lot or batch traceability records; and environmental or shelf life condition records.

One specific requirement concerns electronic records. The organization must be able to show that backup procedures have been defined. Electronic records should be secured to prevent unauthorized alterations or changes and should not be corrupted due to software or system changes.

Findings

This is another common area for findings. Missing records, unreadable records, errors in records, and unexplained gaps in the records have all led to findings. It doesn't take many missing records to elevate the finding to a major finding. Inconsistencies in the control of records or documented procedures, or missing elements—particularly with electronic records—are grounds for findings. The lack of a recovery plan for an information technology disaster is also grounds for a finding under section 4.2.4 and section 6.3.

Having a records storage organization maintain your records constitutes an outsourced process and must be clearly identified and controlled. Ensure that the storage organization is a qualified vendor, because quality records are clearly a process that affects quality.

In AS9120 the computer protocol must show firewalls and system protection to prevent unauthorized changes to records. Failure to do so will result in a finding.

Best Practices

Make sure record retention times are listed in a separate document or as part of the quality manual. Supplier-maintained records are an AS91xx requirement, but ISO 9001:2008's section 4.2.1 has clarified that

> where the term "documented procedure" appears within this International Standard, this means that the procedure is established, documented, implemented, and maintained. A single document may address the requirements for one or more procedures. A requirement for a documented procedure may be covered by more than one document.

Because AS91xx incorporates ISO 9001:2008's requirements, this should apply across the board. The point here is that section 4.2.4's requirement for supplier-maintained records can be located in the purchasing terms and conditions of section 7.4.2. Verify that the retention times imposed on the supplier-maintained records are at least as long as the retention times imposed on you by the customer.

As in section 4.2.3, electronic records must be addressed in the documented procedure or disaster recovery plan for information technology (IT) systems and the AS9120 requirements should be clearly shown. Be sure that all elements of section 4.2.4 (i.e., define the controls needed for the identification, storage, protection, retrieval, retention time, and disposition of records) are covered in the documented procedure. This includes disposition of electronic records and hard copies.

NOTES

Chapter 4
Organization Leadership and Management Responsibility

SECTION 5.1: MANAGEMENT COMMITMENT

Discussion

This section lists five responsibilities of top management that may be delegated but for which top management remains responsible. Consider interviewing top managers to find out what issues they are aware of and to determine their level of commitment.

AS9101D section 4.1.2.2 requires that top management be interviewed at each audit. During this interview the auditor will evaluate the continued relevance of the organization's quality policy and quality objectives; the establishment of performance measures aligned to quality objectives; quality management system (QMS) development, implementation, and continual improvement; top management's commitment to quality objectives; QMS performance and effectiveness; and the organization's performance relative to customer expectations. This will include supplier ratings, scorecard evaluations, audit results, and findings—particularly if there is a history of major or minor findings by the original equipment manufacturers (OEMs). That will be followed up with what actions are being taken to address issues that don't meet customer expectations.

Auditors will be looking for opportunities to corroborate the answers received from managers during these interviews. This includes the availability and relevance of policies and objectives, evidence that these policies and objectives are implemented and understood throughout the organization, determining whether management has direct influence on the continual improvement of the QMS and customer satisfaction, and determining whether top management is involved in management reviews. The absence of objective evidence in this area can be grounds for a nonconformity.

AS9110 requires in this section that the organization establish the safety policy and safety objectives.

Findings

Auditors who are told by top management that implementation of the aerospace quality management system (AQMS) is a "quality issue" outside of management's purview will write findings for lack of management commitment. Although it may be somewhat subjective, the absence of evidence of management review meetings, a lack of management participation in quality objective formulation or tracking, and apparent nonchalance toward customer satisfaction may become major nonconformances.

Note that this is one area in which a finding should be avoided. In fact, any finding related to a clause directed at top management or its duties is to be avoided at all costs. This is because the OEMs will see these findings in the OASIS database and may equate poor performance in the supply chain with top management's indifference.

Under AS9110, failure to establish a safety policy and/or safety objectives results in findings.

Best Practices

It's easy to make a commitment to achieving "best practices," but it's much harder to maintain them. Top management's influence in the development and oversight of the company's AQMS should be readily apparent. Ensure that top management is a required attendee at the management review and that managers are clearly shown in attendance by signature or otherwise. Top managers should be aware of continual improvement activities at the interview. Ensure that they are invited to audit opening and closing meetings and that they discuss audit findings directly with the auditor. It's important to change the focus of attention from just the quality manager to the top management of the organization.

During the interview, make sure that the top manager is aware of each and every one of these requirements. Additionally, make sure the auditor is aware that top management may be much broader than just the individual being interviewed. This may necessitate a second interview with a manager who is aware of the information and data the auditor is seeking in order to satisfy the requirements of section 4.1.2.2. Preparation is imperative, particularly with regard to customer expectations and/or flow-down requirements. For example, if the OEM has written a series of minor or major findings, is the top manager aware of what the findings were, the present status, and the action plan to correct such deficiencies? Another example would be a substandard customer scorecard. The top manager should be aware of the substandard grade and any actions being taken to correct or improve organizational performance.

SECTION 5.2: CUSTOMER FOCUS

Discussion

Top management should know what the customer needs and how the organization satisfies that need. AS91xx requires top management to ensure that product confirmation and on-time delivery performance are measured and that appropriate action is taken if planned results are not achieved. There are a variety of methods available to demonstrate this to the auditor:

- Financial measurement (spreadsheets, forecasts, and business plans)
- Measurement of process performance throughout the organization and at management review
- External measurement, such as benchmarking and third-party evaluation (including registrar audits and customer audits)

- Assessment of customer satisfaction with the organization's products
- Assessment of satisfaction of people in the organization and other interested parties

Findings

Few findings are raised in this section, but those that are included are the result of total indifference to customer requirements or satisfaction and to the AQMS as a whole.

Flow-down requirements from customers are becoming increasingly scrutinized by regulatory agencies. European Aviation Safety Agency (EASA) requirements, for example, should be considered for suppliers. Failure to abide by their contractual obligations will result in findings.

Best Practices

The same practices that apply to section 5.1 apply to section 5.2: Ensure that top management participates at management review, is aware of customer requirements, and considers training and resource requirements and that these results form the basis for the company's objectives.

A new requirement in section 8.2.1, Customer satisfaction, necessitates that the organization track four elements: (1) on-time delivery, (2) quality, (3) customer complaints, and (4) corrective action requests; take corrective actions; and evaluate those corrective actions. Ensure that top management is aware of these requirements, particularly action plans associated with deficiencies in any of the four enumerated elements. Remember that follow-up is an essential element of section 8.2.1. This is directly related to section 5.2, Customer focus.

SECTION 5.3: QUALITY POLICY

Discussion

The quality policy is also called the mission statement. Whatever the name, it must possess certain attributes. Note again that top management is responsible for defining these characteristics.

It's strange that such a small clause creates so many issues at the certification and surveillance audits. Auditors are often informed that the commitment to comply and improve is implied. However, remember the first rule of audit evidence—it must be verifiable. Most auditors understand that despite much training and posting of the quality policy, there will still be employees who seem challenged at describing the quality policy. This requirement can be taken with the fact that the policy statement is not memorized or read from a card. What does the policy mean *to the employee*? That's the important factor.

A review of the quality policy should take place at the management review (per section 5.6.1) and should be recorded in the minutes of the management review meeting. The "appropriate" requirement merely means that it supports the company's needs.

Findings

Findings associated with this section are often the result of employees who seem to know nothing about a quality policy other than "to build good parts." The absence of a commitment to comply with requirements and continually improve within the policy statement is grounds for a finding.

Best Practices

A best practice that would be most helpful here would be the length of the policy statement. Try to keep the policy short enough for the workforce to handle. A full-page quality policy may look impressive, but can it be communicated to and understood by employees? During internal audits, conduct interviews with personnel to verify that they have the required awareness, understanding, and knowledge of how the organization's quality policy relates to their own activities, regardless of the terms they use to express their understanding.

The entire policy statement can be published very simply. If a longer mission statement is needed, separate it from the quality policy statement. The quality policy statement of one organization is usually very similar to all other quality policy statements across the industry. They address compliance with customer requirements, on-time performance, and continual improvement.

SECTION 5.4.1: QUALITY OBJECTIVES

Discussion

Many companies tend to increase the number of quality objectives exponentially to the size of the organization. Larger organizations tend to have objectives scattered at different levels of the organization, and some have excellent tracking programs for these metrics. Smaller organizations, on the other hand, tend to have very few quality objectives. Regardless of their number of objectives, aerospace organizations must ensure that there is a correlation between their quality objectives and their quality policies.

Corporations may flow down their quality objectives to a subsidiary organization. Sometimes these objectives don't fit the quality policy statement or are otherwise unattainable. These objectives are subject to review at management review meetings, and metrics on their progress (or lack thereof) are to be maintained.

Quality objectives need to be measurable and documented. There is no specified way of identifying or documenting quality objectives, as they may appear through business plans, management review outputs, annual budgets, and statements of objectives. However, there should be evidence of the way the quality objectives are passed down throughout the organization's structure and processes.

Findings

This clause is the source of some, but not many, findings. Findings in this area are usually caused by objectives that are impossible to measure, such as "Become the best widget company in the world" or "Be the best company in the world at customer satisfaction." Other common findings are the result of not tracking the metrics that show progress toward the company's objective.

One hidden area concerning this clause is found in section 6.2.2 (d), which requires that the organization

> ensure that its personnel are aware of the relevance and importance of their activities and how they contribute to the achievement of the quality objectives.

This is a common area that needs attention. Some organizations never share the quality objectives with the rank and file, and this omission results in a finding.

Best Practices

Tailor the quality objectives to issues concerning the workforce. "On-time delivery," "quality escapes," "customer rejections," "cost of poor quality," and "customer satisfaction" are good and measurable objectives that should fit into any AS91xx operation.

Section 8.2.1 now requires the organization to track certain parameters. Among these are on-time delivery, quality performance, customer complaints, and corrective action requests. This would be a good spot to also track those items that are required as quality objectives.

Make sure that employees are aware of the objective that affects their specific job. The sales department may not need to be aware of quality escapes, but the production and inspection department should be.

Ensure that metrics are maintained and reviewed at management reviews and that a goal is established for each metric. Start low for a new objective and gradually move the goal to an area that needs improvement. Avoid "100%" as a goal. If you achieve the goal, readjust the objective, remove it, or replace it with another objective. Remember, this is a fluid portion of the standard.

Communicate with corporate headquarters to discuss inappropriate or unattainable objectives. Get corporate's buy-in and correct the objectives for your certificate and AQMS. At least you'll have the paper trail for trying to make it work.

SECTION 5.4.2: QUALITY PLANNING

Discussion

This section is aimed at top management. It references the quality objectives and the imposition of a quality plan, which is defined as a "detailed document that sets forth practices and the sequence of activities aimed at translating an organization's quality policy into operational results, or conformance to a standard such as ISO 9001 within a specified time frame."

When the objectives are set, the top management team has the framework to plan how the organization will achieve them. If the quality objective is set, there should be planning that reflects it. Third-party auditors often use a diagram such as the one in Figure 4.1 to check for this planning.

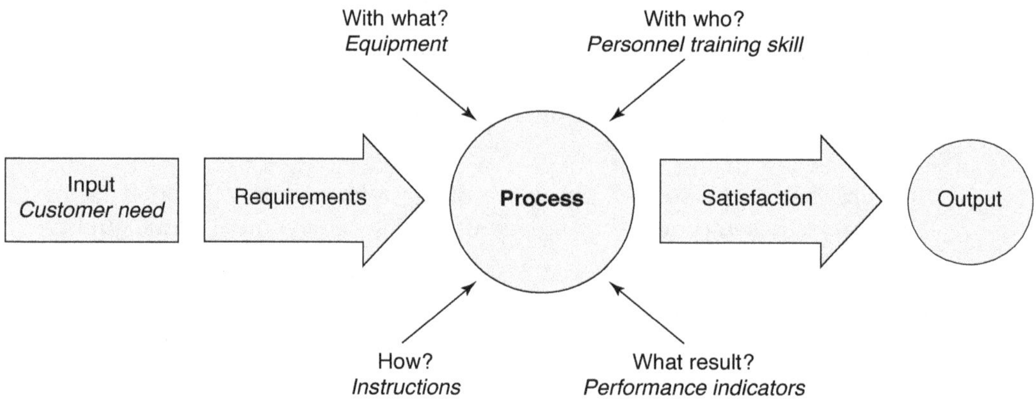

Figure 4.1 Planning process.

Findings

Findings that arise out of this clause are usually the result of a failure to maintain the integrity of the AQMS in some fashion (e.g., planning for short delivery times but not maintaining control of nonconforming products or escapes to the customer). If there are major changes within the organization, there may be no planning to explain or justify the changes. The good news is that there are few findings in this clause.

Best Practices

Ensure that anticipated changes, such as moving a process and its equipment from one location to another, are fully planned with the who, what, and how results documented. New product development should also be planned and documented. Use of a turtle diagram, as shown in Figure 4.1, or a similar tool will help with the planning and its documentation. Make sure the planning tool is always reviewed and documented at the management review meeting.

SECTION 5.4.3: SAFETY OBJECTIVES (AS9110 ONLY)

Discussion

The AS9110 standard assigns top management the task of ensuring that safety objectives, including those needed to meet requirements for product, are established at relevant functions and levels within the organization. The standard further requires that these safety objectives be measurable and consistent with the safety policy.

Findings

This section usually is found in noncompliance if the objectives are either not established at all or, more commonly, established but unmeasurable. A good example is the objective "to have all employees trained to FAA standards" (there is no connection between the safety objectives and any general Federal Aviation Administration [FAA] standard).

Best Practices

Early on, establish a safety policy and objectives and make sure they appear on the management review agenda and minutes for review. Make specific, easily calculated objectives, for example, "fewer than 10 days of lost work per quarter due to OSHA reportable accidents."

SECTION 5.5.1: RESPONSIBILITY AND AUTHORITY

Discussion

Organization charts and job descriptions are the principal elements in this clause.

Findings

Findings usually occur in this clause because there are no clear authorities and responsibilities communicated to the rank-and-file employees.

Best Practices

Develop charts and responsibility matrices that adequately reflect the organization's structure, including lines of authority within departments and also external to them. Ensure clear communication of these tools to all employees at all levels of the organization. This can be accomplished by the use of newsletters, postings, training sessions, and employee orientation.

SECTION 5.5.1.1: ACCOUNTABLE EXECUTIVE MANAGER

Discussion

This position has been labeled "accountable executive manager" specifically by AS9110. The various national aviation authorities have labeled a similar position as necessary for repair stations. Organizations in India and the United States refer to it as the "accountable manager"; aerospace organizations in Canada refer to it as the "person responsible for maintenance."

Whatever the name given, make sure that your AQMS documentation provides for the accountable executive manager in whatever role decided by the authorities.

Findings

Findings generally are found in this section because it is also an FAA requirement.

Best Practices

Develop charts and responsibility matrices that adequately reflect the organization's structure, including lines of authority within departments and also external to them. Include the position of accountable executive manager.

SECTION 5.5.1.2: MAINTENANCE MANAGER

Discussion

The maintenance manager oversees all maintenance services within the repair station. This includes general and specialized maintenance of aircraft and their respective systems, components, and engines. The maintenance manager functions as a leader for the maintenance team within the repair station and is responsible for carrying out the mission, goals, and objectives of the organization. Depending on the organization, the maintenance manager can be the chief inspector, director of maintenance, vice president of maintenance, or some other designation. Make sure that, for AQMS purposes, this individual is clearly defined as the maintenance manager. There may also be more than one maintenance manager.

Findings

This is not an FAA requirement, but it is an AS9110 requirement. There have been findings where the replacement maintenance manager was not identified.

Best Practices

Develop charts and responsibility matrices that adequately reflect the organization's structure, including lines of authority within departments and also external to them. Include the position of maintenance manager or managers and ensure currency of information.

SECTION 5.5.2: MANAGEMENT REPRESENTATIVE

Discussion

In previous versions of ISO 9001, the management representative could be a part-time consultant. However, ISO 9001:2008 states that "top management shall appoint a member of the organization's management." This makes it clear that the management representative must be a company employee and a member of management. There must be clear responsibilities and authority shown for this position that flow from the requirements in section 5.5.2.

The management representative must have the organizational freedom to resolve matters pertaining to quality. This means auditors will look for evidence that the management representative can and does resolve quality issues. The management representative may not be the audit representative for the organization, and there is no requirement that he or she maintain a leadership role, manage, or oversee third-party audits.

Findings

There have been few findings in this area, but the most common occurs when the management representative listed in the quality manual has left the organization and no replacement has been named. Certainly, there would be some follow-up in the form of a finding (requiring a corrective action plan) or some other vehicle to ensure proper replacement. Another common finding involves the organizational freedom requirement; some organizations don't allow the management representative to resolve quality issues without a protracted procedure. Remember that organizations are required to designate an OASIS database administrator. Ask your registrar for details.

Best Practices

Make sure the management representative is clearly identified. If there is a change in personnel, immediately begin—and document—the replacement effort. Appoint an acting management representative for the interim. Some registrars require that a backup be named. Give the management representative freedom to resolve quality issues, and don't burden this authority with convoluted procedures. Clearly show the management representative on the organizational table.

SECTION 5.5.3: INTERNAL COMMUNICATION

Discussion

This section requires that AQMS information be communicated throughout the organization—both up and down the ranks. Note that it only requires communication concerning the "effectiveness" of the QMS. This, again, is aimed at top management. Auditors will look to see what top management has done to communicate quality objectives, the quality policy, management review outcomes, and customer satisfaction data within the organization. It's common for this information to be communicated via bulletin board postings in work areas, which is usually satisfactory. Sometimes, these metrics are continually broadcast on televisions in break rooms, lobbies, and training areas. The information can also be posted on intranets and provided to employees via networked computers.

Findings

There are a few findings in this area and they generally relate to the quality objective requirements included in section 6.2.2. Employees must be aware of the relevance and importance of their activities and how they contribute to the achievement of the organization's quality objectives. If they aren't aware, this is a ripe target for findings. This is especially true where no postings, metrics, or other QMS data can be observed in work or break areas.

Best Practices

Post relevant quality information where employees can easily see it. If you notify them of pertinent quality information in meetings, make sure to document this.

A Microsoft PowerPoint presentation is useful for showing what was communicated. It's important that employees be given all the information necessary to continually help improve the QMS. Newsletters are a good method of communicating quality-related issues, along with employee orientation information, checklists, and work sheets for new employees.

SECTIONS 5.6, 5.6.1, 5.6.2, AND 5.6.3: MANAGEMENT REVIEW

Discussion

Management review is a core required process and one that is verified at each audit, whether certification or surveillance.

ISO 9001 requires top management to review the organization's QMS at planned intervals to ensure its continuing suitability, adequacy, and effectiveness. The review could be performed at a specific meeting, but this is not a specified requirement of the standard. Top management can review the organization's QMS in many ways, such as receiving and reviewing a report generated by the management representative or other personnel, electronic communication, or as part of regular management meetings where issues such as budgets and targets are also discussed. These activities could vary from daily, weekly, or monthly organizational unit meetings to simple discussions or reports.

The required inputs for management review are often misunderstood. The requirements don't limit the review to internal audits. It's expected that the results of all audits—customer, third-party, and regulatory (especially for AS9110 certificate holders)—are addressed. The review of the status of preventive and corrective actions needs to address not just those that were closed but all the actions: what their status is (open, closed, voided) and their effects on the QMS.

One commonly abused input is the follow-up actions from previous management reviews. Make sure all items from the prior meeting that required further action are fully disclosed and discussed. If there were FAA or Transport Canada letters of investigation, corrective actions, compliance notices, or legal action noted at the last audit, these must also be addressed and should be included, in as concise a manner as possible, in the minutes. It would be highly embarrassing, to say the least, for the auditor to know of such activity but not include it in the audit report.

Output from the management review process should include evidence of decisions regarding the three enumerated requirements, but also any decisions regarding changes to the quality policy and objectives, planned and potential process improvements, and revised business plans or budgets. Meeting outputs are also a good source of other important issues, such as plans to introduce new products. Records of management reviews are required, but AS9100 doesn't address the format of these records. Meeting minutes are the most common type of record, but electronic records, Microsoft PowerPoint presentations, e-mails, statistical charts, and metric reports are also acceptable. The management review process should also include the element of QMS planning addressed in section 5.4.2 when new products or changes to existing processes are being considered.

The management review process should not be performed solely to satisfy the requirements of the standard and the auditors; it should be an integral part of the organization's business management process. Half-hearted efforts to satisfy a record-keeping requirement are almost always detected by auditors.

Findings

Findings are common in this section and include a lack of records for meetings and incomplete action items. Lack of a management review is grounds for a major finding. In fact, next to documentation deficiencies, nonconformities related to management review requirements are the most cited source of findings. Incomplete follow-up of prior meeting action items and the status of corrective and preventive actions are also frequent sources of nonconformities. In the latter example, the auditor is often told that there are no corrective actions for a specific area, only to find one when he or she is auditing it.

The management review process is a multifaceted process that's performed at many levels. It should be a two-way process, commenced by top management with input from all segments of the organization. Process activities can vary from daily, weekly, or monthly organizational unit meetings to simple discussions or reports. Auditors will look for evidence that the inputs and outputs of the management review process are relevant to the organization's size and complexity and that they are used to improve the business. Findings will result if they aren't.

If the repair station receives a finding from a regulatory authority, whether for something as simple as a letter of correction to something as complex as certificate action, it should be contained within the management review minutes. Failure to do so would constitute a major finding and put your registration in jeopardy.

Specific to AS9110 is the requirement that the review assess opportunities for improvement and the need for changes to the safety policy and safety objectives. Failure to follow up on the safety policy and objectives is rather common, and there should be no reason to omit this from the review.

AS9110 has three additional requirements for inputs to the management review:

h. results of audits and requests for corrective action from Authorities and customers,

i. the achievement, adequacy and effectiveness of the personnel training program, and

j. changes to Authority requirements that could impact the organization.

These are often missing from the minutes or records of the management review and result in findings.

Best Practices

Create an identifiable management review schedule. It is not necessary, nor is it required, that the management review be handled in a single identified meeting. It can be spread out over the year. As long as it covers all the requirements of section 5.6 it will meet the requirements of the standard. As stated above, AS9110 requires a review of the safety policy and objectives and three additional items.

Top management can review the QMS in many ways, including receiving and reviewing reports generated by the management representative or other personnel within the organization. Electronic communications can be part of regular management meetings in which issues such as budgetary or financial targets are discussed as well as production issues.

Management review is a process. It should be conducted as such and audited internally using the process approach.

Output from the management review process should include evidence of decisions regarding changes to the quality policy and objectives, plans and actions for improvements, changes in resources (whether personnel or infrastructure), and revised business plans.

Auditors will review whether the following points were considered during the management review:

- Will changes to the management system or the business as a whole affect other parts of the system or business?
- Are proposed changes evaluated before implementation?
- Are the requirements of section 4.1 considered when preparing business, strategic, or other processes?
- Are required controls identified before outsourcing begins?

It's important that the management review process not be a hollow exercise performed solely to satisfy a requirement of the standard. Management review should be an integral part of the organization's business management process. Remember that the processes within the standard go two ways: They're generated by top management with inputs from all levels of the organization. This will be reflected when a registrar audits the management review process.

Be sure to include any production meetings, sales meetings, or department meetings in the schedule and make the records of these meetings readily available. Remember that production meetings and periodic management meetings can be part of a management review. Just take care to include them in your procedures of discussions during the audit. Keep minutes of these meetings if they are part of the management review process. Create an agenda that covers all of the input (section 5.6.2) and the output (section 5.6.3) requirements. It's a good idea to keep hard copies of all the minutes, metrics, and inputs/outputs in a folder so you can show the auditor that all items have been covered—regardless of the meeting frequency, content, and attendees. Be sure that top management is shown to participate. Finally, make sure you have a full-disclosure policy with regard to authority action.

SECTION 5.7: SAFETY POLICY (AS9110 ONLY)

Discussion

In addition to the section 5.3 requirement for the organization to establish a quality policy, AS9110 requires a safety policy. Note again that top management is responsible for defining these characteristics:

> Top management shall ensure that the safety policy is appropriate to the purpose of the organization, and that it includes a commitment to comply with requirements and continual safety improvement. In addition it must provide a framework for establishing and reviewing safety objectives and that is communicated and understood within the organization, and is reviewed for continuing suitability.

Just as with section 5.3, it's strange that such a small clause creates so many issues at the certification and surveillance audits. A review of the safety policy should take place at the management review (per section 5.6.1) and should be recorded in the minutes of the management review meeting. The "appropriate" requirement merely means that it supports the company's needs.

Findings

Findings associated with this section are often the result of employees who seem to know nothing at all about a safety policy. The absence of a commitment to comply with requirements and continually improve within the policy statement is grounds for a finding.

Best Practices

Again, a best practice that would be most helpful here would be the length of the policy statement. Try to keep the policy short enough for the workforce to handle. A full-page quality policy may look impressive, but can it be communicated to and understood by employees? During internal audits, conduct interviews with personnel to verify that they have the required awareness, understanding, and knowledge of how the organization's safety policy relates to their own activities, regardless of the terms they use to express their understanding.

NOTES

NOTES

Chapter 5
Human and Other Resources

SECTION 6.1: PROVISION OF RESOURCES

Discussion

To fulfill the customer's expectations, it's incumbent on the organization to allocate enough resources to meet delivery schedules and quality requirements, and to continually improve the processes. This responsibility isn't limited to human resources; it applies to all the resources it takes to create the deliverable, including materials, infrastructure, and utilities. This means that appropriate resources are to be identified, planned, made available, used, monitored, and changed as necessary by the organization. The organization should verify whether it has evaluated past and present performance (e.g., using cost-benefit analysis and/or risk assessment) when deciding what resources are to be allocated. Outputs from the management review should focus on this element and document the organization's efforts in analysis. AS9110 has a requirement that the organization have a system to continually assess the availability of tools, technical data, and necessary qualified personnel in order to safely complete maintenance, repair, and overhaul activities.

Findings

Findings are common in this section when layoffs or workforce reductions result in systemic repeats of delivery, quality, and customer problems. A sudden increase in customer complaints and rejections, on-time delivery slipping by increasing percentages, and even a sudden decrease in overall customer satisfaction are all systemic evidence of resource deficiencies. Failure to include the continual assessment for AS9110 (say "internal audit") also results in findings.

Best Practices

Be sure to link the resource management processes to the management review. In fact, the following processes should be treated as continuous:

- Management commitment and responsibilities
- Management review process
- Internal audit activities for AS9110

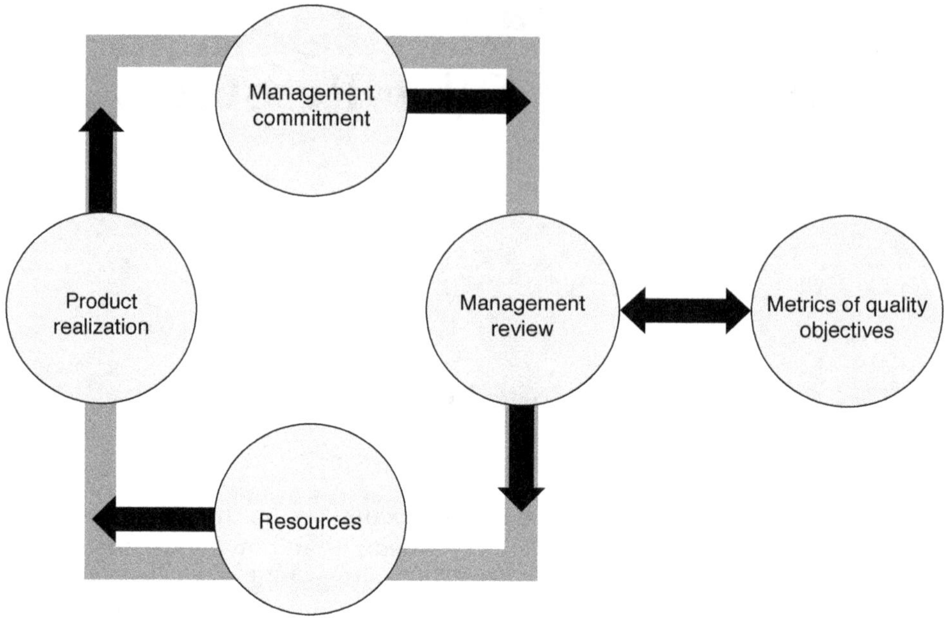

Figure 5.1 Linking quality objectives.

Ensure that human resources (6.2), infrastructure (6.3: energy, water, facilities; and equipment maintenance, communications, information technology), and the work environment (6.4: temperature, lighting, vibration, noise) are adequate and maintained in a fashion that conforms to product requirements, as shown in Figure 5.1.

SECTIONS 6.2 AND 6.2.1: HUMAN RESOURCES

Discussion

This is a simple requirement that employees be competent in the performance of their duties. Note, however, that the basis of competency includes four elements: education, training, skills, and experience. These four elements will be repeated in the next section. This short section has big implications. There are two kinds of qualifications that the organization must address for any position:

- *Required qualifications.* This is the specification of the required experience of the employee, skills he or she must possess, educational requirements of the employee, and required training, whether given by the organization or other sources (e.g., prior employment, technical schools, regulatory agencies, original equipment manufacturer [OEM], or the military).
- *Attained qualifications.* This is the sum and substance of what the worker has acquired at the present time.

AS91xx states that employees' attained qualifications must equal or exceed required qualifications. This puts pressure on the human resources department to maintain job description requirements that meet product realization requirements without requiring employees to be overqualified. In AS9110, noncertified personnel must be assessed prior to performing work, so you must devise a methodology for evaluating noncertified personnel and maintain a record of such evaluation.

Findings

Findings are routinely written when the organization cannot show how positions are filled and with which qualifications. Another common problem area is where the job description for a management position specifies education requirements and the employee doesn't have them (e.g., the job description states that the purchasing manager must have a degree in business or finance, and the manager doesn't have one). Finally, ensure that there are identifiable qualifications to become an internal auditor.

Best Practices

Undoubtedly, the best practices for section 6.2.1 are closely linked to those in section 6.2.2. A training matrix is a simple tool that will assist in identifying them. There are many variations, and the sample shown in Figure 5.2 illustrates one of them.

The point of this section is to note that training and qualifications should match the job description for inspectors, that *attained* qualifications equal the *required* qualifications. A variation of the chart in Figure 5.2 could be used for each department, but small organizations may be able to list all employees and positions on one chart. Make it work for you. Don't overwrite job descriptions or require skills that cannot be verified.

Finally, review your matrix with your team regularly and according to changing circumstances. Update it as needed; it's a dynamic document that may change due to changes in priority, personnel, or the time of year. Reviewing and updating the matrix should be discussed and recorded during management review.

SECTION 6.2.2: COMPETENCE, TRAINING, AND AWARENESS

Discussion

In keeping with the process approach, employee competence, training, and awareness should be audited as part of a product realization process audit, not as a separate process. Most—but not all—organizations have separate human resource processes where most of the evidence needed can be found.

Boogaloo Bearings Inc. — **Certified FAA Repair Station No. 12CR2409**

Chief Inspector J. Munsen
1234 Home Ave.
Salmon, ID

Inspector name	General	Nondestructive inspection	Component maintenance manual	Travels	Use of micrometers	Bore gages	Use of height gages	Dal indicators	Calipers	Illustrated parts and manuals	Company forms/ manuals	Other

Figure 5.2 One-page training matrix.

To satisfy the competence/effectiveness requirements of an AS91xx-compliant aerospace quality management system (AQMS), an organization will typically need to do a variety of things:

- Identify which competencies are required by personnel performing work that affects quality
- Identify which personnel already performing the work have the required competencies
- Decide which additional competencies are required, as well as recurrent training requirements
- Decide how these additional competencies are to be obtained: training of personnel (external or internal), theoretical or practical training, hiring of new personnel, or assignment of existing competent personnel to different work functions
- Train, hire, or reassign personnel
- Review the effectiveness of actions taken to satisfy competence needs
- Periodically review competence of personnel and conduct recurring training as required
- Maintain appropriate records of education, training, skills, and experience

The standard is silent on the exact nature of the records to be maintained. On-the-job training (OJT) records meet the same requirements as other records under this section. In auditing an organization's compliance with the competence and training evaluation requirements, an auditor will look at OJT records as well as the more traditional training records.

The organization must have a systematic approach to identify worker competencies and to verify that its approach is effective. The outcome of the process may be a spreadsheet, roster, database, human resources plan, competencies development plan, contract, project, or product plan. The organization must also maintain records required by the appropriate regulatory authorities.

New or changed activities or processes that may lead to different competency requirements should be evaluated. In AS9110 this might include new aircraft, engines, accessories, or specialized tooling, such as cold forming riveting. A review of competencies might also be needed when a new tender or contract is being considered. Evidence of this should be found in the management review outputs and minutes. Competence requirements may also be included in contract documents when the activities of workers, outsourced processes, and subcontractors can have an effect on processes and/or product quality characteristics.

The organization should evaluate the effectiveness of the actions it takes to satisfy competence requirements. This is an important, yet often overlooked, element of implementation. The organization may use any number of techniques, including testing, peer or supervisor review and/or observation, reviews of training and employment records, and interviews.

Table 5.1 contains examples of the methodologies of evaluation.

Training records should verify that a training course has been successfully completed, but note that this alone will not provide evidence that the trainee is competent. Mere attendance of a training course isn't evidence of evaluation or competence.

Table 5.1 Competency evaluation methods.

Evaluation method	Objective	Examples
Review of records	To verify the background of the auditor	Analysis of records of education, training, employment, and audit experience
Positive and negative feedback	To provide information about how the performance of the auditor is perceived	Surveys, questionnaires, personal references, testimonials, complaints, performance evaluation, peer review
Interview	To evaluate personal attributes and communication skills, to verify information and test knowledge, and to acquire additional information	Face-to-face and telephone interviews
Observation	To evaluate personal attributes and the ability to apply knowledge and skills	Role playing, witnessed audits, on-the-job performance
Testing	To evaluate personal attributes, knowledge and skills, and their application	Oral and written exams, psychometric testing
Post-audit review	To provide information where direct observation may not be possible or appropriate	Review of the audit report and discussion with the audit client, auditee, colleagues, and the auditor

The organization must demonstrate that its personnel have attained competence through a combination of education, training, and/or work experience.

The organization must verify and show that some form of an effective monitoring process is in place and being acted on. Ways of doing this include a continuing professional development process (such as the one described in ISO 19011); regular appraisals of employees and their performance; and the regular inspection, testing, or auditing of product for which individuals or groups are responsible. Ongoing changes in competence requirements may indicate that an organization is proactive in maintaining personnel performance levels. Recurrent training requirements are required in AS9110 and by many regulatory agencies (even for AS9100 compliance). Any recurrent training requirements set by the organization for compliance with AS91xx should be clearly defined. A good example of this is the human factors training requirement imposed by both AS9110 and the European Aviation Safety Agency (EASA).

Findings

This is a common area for findings in all AS91xx-compliant organizations. Many findings are written for lack of training records, recurrent training requirements, and establishing necessary employee competencies. AS9110 registrants are often cited for failure to provide training and for late or undocumented recurrent training. Orientation training that provides a foundation for new employees on the company's philosophy, procedures, and processes is usually mandated but not always given. When it is provided, it may not be updated or recorded. Frequently, OJT records—or the lack of them—lead to findings. This is an area that tends to be overlooked in implementing and auditing a quality management system (QMS). The previously discussed area of worker awareness of how their activities contribute to achievement of the company's quality objectives is another common source of findings. Failure to train temporary or part-time employees is also a common finding.

Best Practices

Remember that this clause is usually audited as part of a process audit. Make sure that you generate a checklist for all orientation training given, even to part-time and temporary employees, and that you have some methodology to evaluate this training. Simple true-false tests are a sufficient answer to many issues concerning evaluation. If there is an observation evaluation, be sure to record it. To avoid creating huge numbers of paper evaluations, consider an individual training matrix record, such as the one illustrated in Figure 5.3. Transcribe the necessary information and discard individual records.

Training matrix for: ______________________________
(name of employee)

Date	Subject	Instructor	Duration	Score/remarks

Individual training record ITR-1 Rev 2/4/14

Instructions: The instructor will enter the subject taught, the date given, and the duration of any training given to any employee. The remarks section can be used to enter any other information deemed important by the instructor or chief inspector.

Figure 5.3 Training matrix.

Periodic newsletters issued to the workforce are another great source of training material, as long as the organization maintains records of their distribution. Organizations, regardless of size, should lean toward online records and self-taught courses with evaluations embedded in the training. Make sure that job-specific items, such as electrostatic discharge, are covered in training sessions.

Centralize the records as much as practicable. Scattered records often result in unnecessary findings. Make sure that any personal external record that an employee uses for establishing competency is included in his or her training record at your organization. These records include former military experience, schools, and related experience. One word of caution regarding certified operator programs (COPs): There are many COP-related findings because organizations don't provide proper or follow-up training or maintain records for it. Have COPs stand in the shoes of an inspector. Verification of inspection points is a very sensitive requirement of AS9100C.

SECTION 6.3: INFRASTRUCTURE

Discussion

Often minimized, this section frequently has a sizable number of findings during audits. A change in the 2008 update to ISO 9001 (that is applicable to aerospace organizations) is the inclusion of information technology in section 6.3: "The organization must 'determine' the infrastructure requirements." This requires the organization to address how it determines the infrastructure necessary for achieving effective and efficient product realization. The registrar will look at whether a plan is in place for the maintenance of the infrastructure and the scope of the plan (i.e., workspace, equipment, and support services such as information technology, transport facilities, and communication). An additional requirement in AS9110 requires planning for suitable facilities, acceptable to applicable customers and authorities, for performing maintenance, repair, and overhaul services away from the organization's fixed location.

Findings

Specification-required infrastructure (e.g., humidity, temperature) criteria must be met. Findings arise when recording devices for these elements aren't calibrated or are nonexistent. Failure to possess a disaster recovery plan for IT (especially if the records required by clause 4.2.4 are maintained on the servers) is grounds for a finding. Missing required tooling, fixtures, and equipment also results in findings, some of which may be major findings if the omissions affect the quality of the final product or product realization. "Brown-out" protection or alarm devices may be necessary on electrical tooling and equipment in areas affected by electrical supply disturbances; failure to install these may lead to findings. Failure to show any planning for "away from base maintenance" under AS9110 will also result in a finding.

Best Practices

Make sure in your quality planning from section 5.4.2 to analyze and accommodate for possible infrastructure shortcomings and needed tooling, fixtures, and equipment. This should, as always, appear in the records of the management review. A common recurring issue has been to make sure that the IT plans, including the disaster recovery plan, are controlled under section 4.2.3. They affect the quality of the final product and product realization.

SECTION 6.4: WORK ENVIRONMENT

Discussion

Safety and the handling of the environmental requirements of the product are the focus of this section. The primary purpose of this area's audit is to ensure that the organization determines and manages the work environment to achieve conformity to product requirements. The organization should have established safety procedures, communicated them to all employees, and provided proper safety equipment, such as safety glasses. Auditors will check to ensure that the organization's facilities have proper heat, humidity, light, and airflow to effectively accomplish its production or servicing and that the work areas are clean. This section may be used in conjunction with section 7.5.1.

Findings

Findings may be written against section 6.4 or section 7.5.1. Missing or unchecked electrostatic discharge equipment is a common finding. Violations of foreign objects and debris (FOD) procedures often find their way into minor and major findings. Another common finding concerns freezers that store temperature-sensitive and time-sensitive materials that have no recording devices to verify that their temperature limits have been maintained. There will be increased scrutiny of FOD programs in the future, and OEMs often have specialized FOD training and assessment programs.

Best Practices

Make sure that any equipment requirements are fully evaluated and that the equipment has the correct, calibrated component necessary to prove compliance with requirements. Make sure work areas are clean. Always put your best foot forward on audit day, as Industry-Controlled Other Party (ICOP) personnel may accompany registration auditors (think OEM). Document any problems that arise on preventive maintenance issues so that you can show that your organization strives to maintain the proper work environment for product realization.

Additional training may not be required if personnel are already competent. Auditors have already been loosely applying this configuration. For example, a newly hired certified welder doesn't need training just for the sake of the training requirements of AS9100C. Train only when it's necessary. The phrase "satisfy these needs" in earlier versions of AS9100 was terribly unclear and needed clarification.

The placement of IT systems in this section indicates that the electronic age has hit the AQMS. Section 6.4 requires the implementation of disaster recovery plans, backup plans, and similar plans for the auditor to review. The rationale for making this change is the "insurance" it provides the records and programs already required by clause 4.2.4. This will take additional effort, but most companies already back up their data and have recovery plans.

NOTES

Chapter 6

Product Realization and Customer-Related Processes

SECTION 7.1: PLANNING OF PRODUCT REALIZATION

Discussion

Section 7.1 has one new aerospace requirement, and it represents a continuation of the theme that the organization must identify resources to support operation and maintenance of its product. As shown throughout sections 4 to 6, there are clear requirements for organizations' products or services regarding the following:

- Preparation and tracking of quality objectives
- Determination of the requirements for the organization's products or services
- Documentation requirements
- Analysis of resource needs
- Need for maintenance of the requirements for the products or services

Sections 7 and 8 require even more product realization steps:

- Design and development actions
- Assembly and manufacturing processes
- Inspection requirements
- Evidence of conformity

Together, these activities form part of a quality plan. The standard requires that a document specifying the processes of the quality management system (QMS)—including product realization processes—and the resources to be applied to a specific product, project, or contract be in a form suitable for the organization's method of operations. A flowchart illustrating the activities related to this process is shown in Figure 6.1.

This plan should address the processes that will touch the product or service that requires compliance. AS9110's requirements concern servicing and will

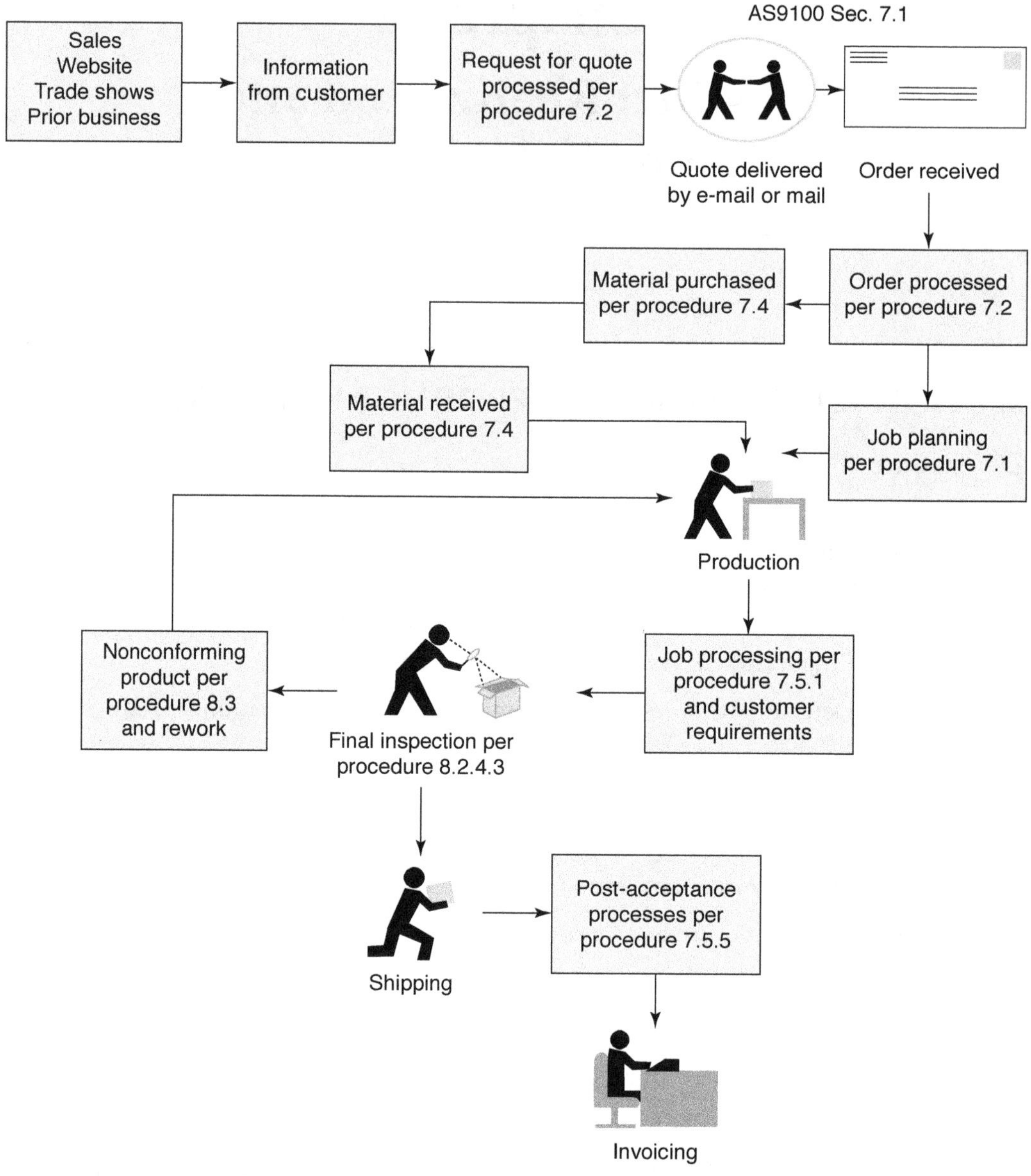

Figure 6.1 Product realization plan.

usually be shorter due to the simplified regulatory requirements for the processes. AS9110 has a specific planning note:

> NOTE: Quality objectives and requirements for the product include consideration of aspects such as
>
> - product and personal safety,
> - reliability, availability and maintainability,
> - producibility and inspectability,

- suitability of parts and materials used in the product,
- foreign object debris/damage (FOD),
- selection and development of embedded software, and
- recycling or final disposal of the product at the end of its life.

Likewise, AS9110 also requires that planning include configuration management appropriate to the product; the identification of resources to support the use and maintenance of the product, in order to keep it airworthy; and safety objectives and requirements for the product.

ISO 9001:2008 added measurement as one of the required activities to be determined during the planning of product realization. One of the most significant changes in AS91xx is the move of configuration management from section 4.3 to section 7.1.3. Accordingly, aerospace auditors won't address configuration management as a stand-alone process but as part of product realization. This is an improved method of handling this critical aerospace requirement.

Findings

Findings in this area usually stem from a process audit that followed the flowcharted path for compliance with the points specified when an intermediate process that was required was skipped or not recorded. AS9110 requirements are also scrutinized for compliance with the specific planning requirements; if the organization cannot show the requisite planning, a minor or major finding may be issued. Examples might be engine or exhaust plugs to prevent FOD, pitot head covers, or safety equipment (personal protection equipment or fire extinguishers).

Best Practices

This seems to be an area that confuses many organizations. Map your process from the customer order to the production department. Have a checklist for the AS9110 requirements. The focus of this section of the standard is to simplify the process. The more steps the process has, the greater the chance that one of them will be missed or undocumented. Reduce the number of required approvals to only those necessary to fulfill the requirements of your process. For example, don't require 15 signatures to approve a contract. The chances that those 15 people will be present each and every time you need them can't be guaranteed. If numerous signatures are required, make sure you have a mechanism to get all of them.

To maximize efficiency, consider combining this process with your efforts to comply with the requirements in section 7.2. This is illustrated in Figure 6.1.

SECTION 7.1.1: PROJECT MANAGEMENT

Discussion

AS91xx includes a new requirement for planning and managing product realization in a structured and controlled way to meet requirements within acceptable risks and within resource and schedule constraints. Project management is expected for all products at all levels of the supply chain. Auditors will ensure that the organization has a defined process for managing product realization to ensure quality and that the schedule is not compromised. This requires the organization to plan product realization on the basis of project management. Depending on the complexity of the product, this may be as simple as a shop traveler or as elaborate as an entire project management process and procedure. Section 7.1.1 introduces the major change in AS91xx that's included in 7.1.2: the inclusion of risk analysis as part of the aerospace quality management system (AQMS).

Findings

The new requirements for project management dictate that planning be performed in a structured and controlled way. If there are no documents, procedures, processes, guidance, or forms in place that identify at least the methodology and control of the project management, a finding will be issued. Likewise, failure to conform to project management requirements is also a finding.

Best Practices

Unless you are a very large organization, this shouldn't be too difficult a task. Most small and medium-sized manufacturers and repair shops use work orders or travelers for controlling projects. Just make sure that you identify this methodology somewhere in the process. If you are using a hybrid system that calls for travelers to be used for projects under a certain dollar amount, make sure you have a written process on when and how the project manager may be used for projects that exceed that dollar amount. Make sure that this aligns with section 7.1.2, Risk management.

SECTION 7.1.2: RISK MANAGEMENT

Discussion

In AS9100B, risk was addressed only in clause 7.2.2 (d) to review product requirements for specified risks, such as new technology and short delivery times. AS91xx's expanded focus on risk includes risks posed during product realization planning, which requires more effort for organizations.

Findings

Section 7.1.2 contains the major changes to AS91xx and has created many findings since audits to revision C began. The failure to identify a risk program poses the risk of a major finding. Remember that this section, like sections 7.1.1 and 7.1.3, contains the modifying phrase "as applicable." This means that the amount of risk management in your organization depends on its size and complexity, among other factors. Create a risk program that follows the flow of product through your system. Most manufacturing companies consider risk during the quoting stage. Follow the risk mitigation instructions generated for your QMS.

Best Practices

Risk analysis should be performed as part of any contract negotiation, contract edition, or contract modification. Data used in risk management should be linked to potential issues and problems before they can cause cost and/or scheduling problems. In larger companies, risk management can be quite involved. In those cases, NASA work instruction 170846 provides an excellent description of risk analysis.

A risk analysis can be depicted on a process map, such as the one depicted in Figure 6.2. Numerous publications detail what risk assessment is and how it's calculated. Many of these publications originated in the federal government's space exploration programs and the related prime contractors. The risk management program required by AS91xx section 7.1.3 doesn't have to be as detailed as that for a shuttle launch. Aerospace organizations must simply analyze their ability to deliver products and services and what factors may impede delivery. The standard requires that any risk management program contain the following:

- Assignment of responsibilities for risk management
- Definition of risk criteria (e.g., likelihood, consequences, risk acceptance)
- Identification, assessment, and communication of risks throughout product realization
- Identification, implementation, and management of actions to mitigate risks that exceed defined risk acceptance criteria
- Acceptance of remaining risks after implementing risk mitigating actions

On the other hand, a small machine shop or repair station may have only three apparent risks:

- Can a job be done on time?
- Is the material available?
- Is there enough equipment and workers to perform the job?

The risk analysis required for such projects (see "Discussion" in section 7.1.1) is considerably less for a small company than for a larger company. Many jobs are repetitive, and so a risk analysis wouldn't be required. Set a trigger point for when a risk analysis must be performed. Bear in mind that this doesn't absolve the organization from the need for a risk mitigation program. Refer to Appendix C for an example of a small organization's risk assessment program for a quoting procedure. The bottom line is that section 7.1.2 requires a written process to document an organization's risk management process.

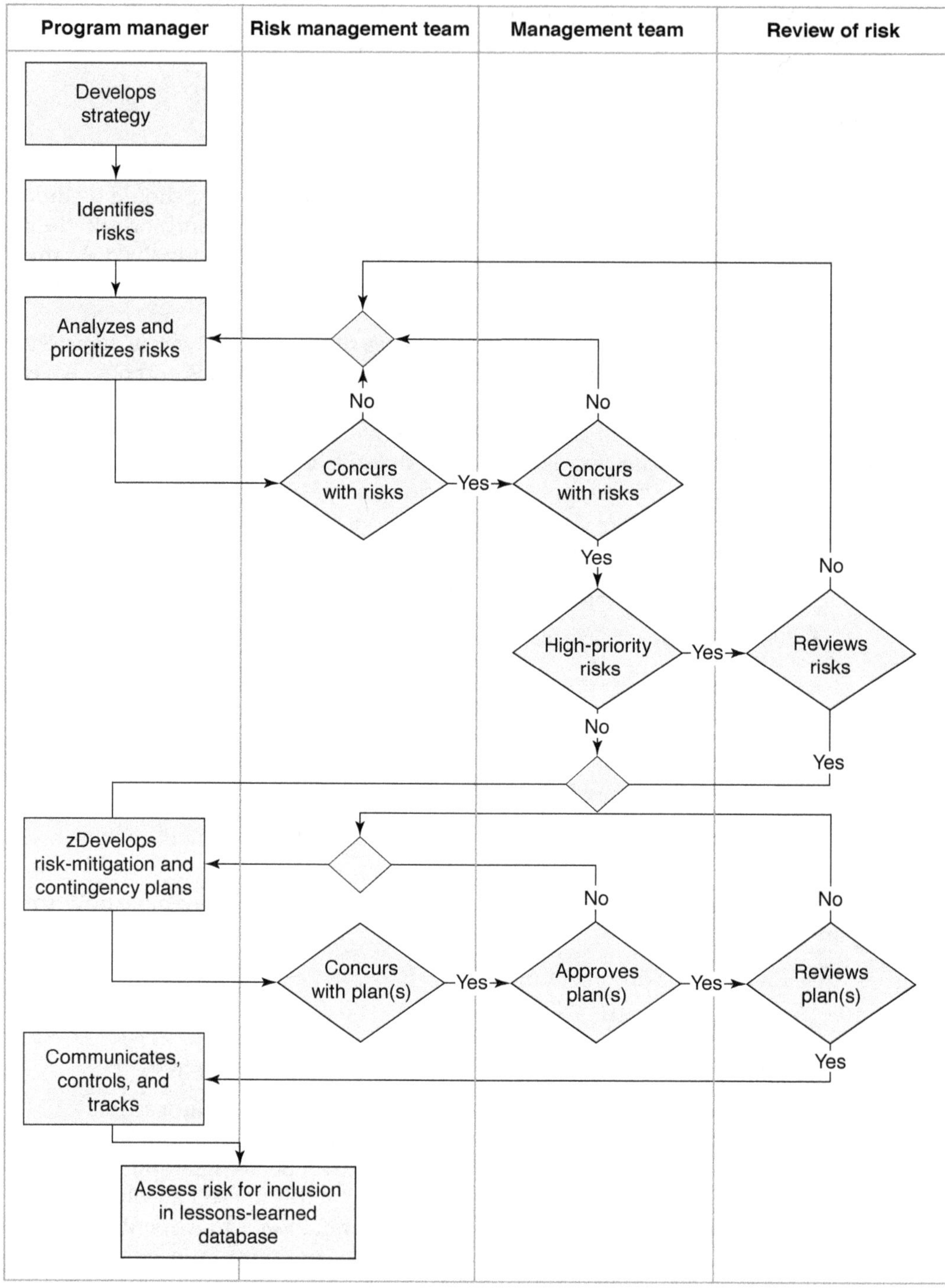

Figure 6.2 Risk analysis matrix.

SECTION 7.1.3: CONFIGURATION MANAGEMENT

Discussion

Document control is a very large part of configuration management. The configuration management plan establishes and maintains consistency in product performance and manufacture. It also records a product's functional and physical attributes, material requirements, design, and maintenance.

Organizations must ask, "Can the product be configured into different products, and can they exist at different revision levels simultaneously?" Some components on original equipment manufacturer (OEM) aircraft, engines, propellers, and appliances can be placed on several models or products, each with a different part number or variation number. An example would be a reciprocating powered engine designated as O-360. Depending on the use of the engine, it may be an O-360 L1A, O-360-G2, or I/O 360-H1. This would require a rather refined configuration management system, which is required by AS9100C. ISO 10007 provides excellent guidance in this process.

The International Aerospace Quality Group (IAQG) believes that the proper configuration management implementation can positively affect product quality and on-time product delivery, and decrease development and life cycle costs. It suggests that all discrete parts, assemblies, and units be identified by part numbers in accordance with the applicable standards and/or customer definitions. Serial and lot numbers should be permanently and uniquely identified by a discrete serial or lot number. There should be coordinated activities for controlling change to configuration baselines. Configuration control should be applied to establish baselines, configuration identifications, and all released engineering data. The configuration control system should stream, classify, and process changes so that it provides fully coordinated and approved change packages in compliance with the requirements of the contract.

AS91xx's previous revision made configuration management almost a footnote. Its section 4.3 required only that organizations have a configuration management process appropriate to the product. Section 7.1.3 of AS91xx is considerably more rigid. For organizations that exclude design and development under section 7.3, the exclusion does not affect this requirement.

Some of the new requirements include a configuration audit and configuration status accounting. All of this information is contained in ISO 10007. Be ready to explain to the auditor how your organization maintains the configuration management of the product from the time of the purchase order through delivery. Generally speaking, organizations can explain that final inspection constitutes the configuration audit. Make sure that such inspection deals with the configuration of the product.

Findings

Auditors will look for the information that is used to produce a specific configuration of the product. Deviation is usually grounds for a major finding, as product quality will be affected and, in these cases, recall of product might be required.

This information includes the revision levels of constituent parts and subassemblies. Improper exclusion of this requirement is also a major finding.

Change control for configuration changes must be strictly maintained or a finding will be issued.

Best Practices

Organizations that assemble components using manufacturers' specifications, drawings, parts lists, or bills of material must ensure a continuity of flow and that all revision levels are current.

Manufactured products made from scratch require the engineering department to ensure that the configuration of the product is what the organization ordered.

Make sure that the configuration management program applies appropriate processes and tools to establish and maintain consistency between the product and the product requirements and attributes defined in the product's configuration information. Confirm that products conform to their requirements and are identified and documented in sufficient detail to support the product life cycle.

The internal audit program should apply surveillance to identify and document the functional and physical characteristics of a configuration item, to verify control changes to those characteristics, and to verify recording and maintenance of changes to processing and the implementation status of various products. Far too little internal audit activity is often noted in this area. AS91xx references "configuration audits" and "status accounting." Make sure to do periodic checks on product configuration as part of the internal audit program.

SECTION 7.1.4: CONTROL OF WORK TRANSFERS

Discussion

It's often necessary to temporarily outsource work. This may be the result of special processes that an organization doesn't provide or that are required by a customer. Regardless of the reason for outsourcing, the quality of any outsourced product must be controlled and validated in a defined manner. Section 7.1.4 aligns with the requirements for outsourcing in ISO 9001:2008 and now applies to both permanent and temporary transfers.

Findings

Findings in this area are usually the result of a discrepancy between a requested product and a received product, missing or incomplete documentation, or missing test results. The receiving inspector is usually charged with verifying products as they are received into the organization.

Best Practices

The organization must have a procedure for inspecting outsourced materials. For example, if a component is sent out for chrome plating, the organization must ensure that the returned product includes a certificate of conformity to the exact process requested on the purchase order. If the request was for a Mil-Spec and the certificate includes a representation of performance to an ASTM (American Society for Testing and Materials) specification, the receiving inspector should catch this discrepancy and handle it accordingly.

AS91xx's requirement for control of work transferred outside the organization has been moved to section 7.1.4 in AS91xx and still applies. In its place, post-delivery support has been added at section 7.5.1.4.

SECTION 7.2: CUSTOMER-RELATED PROCESSES

Discussion

Section 7.2 is broken down into three distinct parts: Section 7.2.1 applies to the processes in which the product or service to be provided are determined, section 7.2.2 applies to the product or service to be provided, and section 7.2.3 applies to communication between the organization and the customer. Section 7.2.2 contains the only deviation from ISO 9001 by incorporating risk analysis into this process, which is commonly referred to as contract review.

The ability of any organization to gain market share and profit depends on its adaptability, corporate culture, relationship with stakeholders, and capability to continually improve. Risk management is a concept finding its way into processes that support sustained improvement and achievement.

Section 7.2.1 requires organizations to look beyond their customers' stated requirements into what they really *need*. Registrars will expect that all customer needs—stated and unstated—as well as the organization's knowledge of potential uses of the product will be examined. Any regulatory agency requirements (such as the Transport Canada Civil Aviation [TCCA] or FAA for AS9110-registered organizations) shall be considered and accommodated.

Findings

When planning for a new product or service, an organization has a responsibility to accommodate statutory and regulatory requirements. Findings are noted where these inquiries aren't performed or documented. An example would be a toy manufacturer that uses materials known to be toxic. Another requirement is the "necessary for specified or intended use, where known." Organizations that know their products' intended uses are required to account for these details. Findings have been issued when organizations ignored or did not document the details.

Best Practices

Registrars don't know the ins and outs of their clients' businesses as well as their clients do—at least they shouldn't. Thus, they might use checklists to ensure that their auditors examine everything they should examine during an audit. This is a theme that is repeated throughout sections 7.2.1 and 7.2.2. Checklists should include the requirements of the applicable regulatory bodies. These include the

Federal Aviation Administration (FAA), US Environmental Protection Agency (EPA), US Food and Drug Administration (FDA), Occupational Safety and Health Administration (OSHA), and NASA at the federal level, and transportation departments and other state agencies at the state level. Professional and industry-specific requirements should also be included, if appropriate. Make sure that required procedures and practices are in alignment. This should be a focus area of the internal audit to discover evidence of misalignment.

SECTION 7.2.2: REVIEW OF REQUIREMENTS RELATED TO THE PRODUCT

Discussion

This section focuses on the product rather than on processes. There are several factors to consider in the requirement. This section applies to the original contract and any changes to that contract, both of which should be documented in the contract review procedure developed to manage this process.

The note in this section amounts to a huge escape valve. If the organization has a large internet sales presence, it should be detailed in the procedures. Here, the practicality is that the offer to sell on a website is different from a request for quote or a request for proposal from a customer, which requires intense negotiation or review. The auditor will review the internet information, advertising literature, and catalogs, along with the procedure developed to manage this process.

The standard requires that records of the review and any resulting actions be maintained. The main addition for the aerospace industry concerns risk assessment. Risk analysis should be part of every contract negotiation. It will help identify potential problems before they can pose cost and/or schedule issues. It forces aerospace companies to ask themselves, "Can I produce it in the time frame requested by the customer, and can I get the materials that I need? Do I have the trained workforce and equipment necessary?" This is the analysis necessary to conduct business in the first place. The key is that this analysis must be made and documented prior to acceptance of the job.

AS9110 also has a specific requirement that contractual requirements be reviewed so that special requirements of the product can be determined (e.g., scope of work, technical data, delivery requirements, requirements regarding subcontracting of work).

The AS9100-specific requirement for review of maintenance contracts should exist within the organization. Ensure that the contract review includes a definition of requirements regarding subcontracting of the work. Finally, some regulatory authorities, such as the European Aviation Safety Agency (EASA), have embedded requirements concerning the use of Parts Manufacturer Approval (PMA) parts. The registrar will want to know whether the aerospace company has written permission to use them.

Contract amendments must meet the same requirements as the original review. AS9100 requires that the review on amendments include provisions for disposition of out-of-scope defects discovered during maintenance. In fact, the common practice of submitting components for "evaluation" and then quoting bears on this area.

Findings

Findings occur with some frequency in this area, often when records of the reviews are not available. Contracts that are accepted without review or with a review after acceptance are grounds for a finding. When there are unresolved differences on current contracts between the organization and the customer, a finding will be issued. AS9110 contracts need to discuss "out-of-scope repairs" or a finding will be issued.

Best Practices

As previously stated, a checklist is the best protection from variance from the standard's requirements. Make sure that this review occurs prior to acceptance of the contract. If the organization performs internet sales, a full review isn't possible or necessary, but the procedures should make the process clear—especially if internet and traditional sales methods are employed.

The review should be tailored to the organization's business and the usual process for contracting. For example, an AS9100-registered firm that performs the same tasks for the same customers may not even know what jobs will be done until they show up at the loading dock. In that case, the receiving inspector may perform the review to determine whether the organization can complete the process in a timely manner. As long as the review and risk assessment are done prior to acceptance, the requirement should be satisfied.

Section 7.2.2 has a close ally in section 7.2.3, Customer communication. Ensure that contract terms, statements of work, requests for quotes, requests for proposals, clarifying language, and changes are maintained as part of this process. Changes may be simple, such as delivery dates and quality increases, as well as major, such as design changes. Ensure that your change procedure is in alignment with the process. See the discussion in section 7.1.2, Risk management. Historically, risk has always been considered in section 7.2.2. And while it is still discussed in section 7.2.2, the new requirements in section 7.1.2 are more important.

Make sure your company has a clear policy regarding the use of PMA parts and has permission to use PMA parts on aircraft and equipment whose regulatory authorities require such permission.

Section 3.2, Special requirements, aligns with the risk management requirements in section 7.1.2:

> Those requirements identified by the customer, or determined by the organization, which have high risks to being achieved, thus requiring their inclusion in the risk management process. Factors used in the determination of special requirements include product or process complexity, past experience, and product or process maturity. Examples of special requirements include performance requirements imposed by the customer that are at the limit of the industry's capability, or requirements determined by the organization to be at the limit of its technical or process capabilities.

SECTION 7.2.3: CUSTOMER COMMUNICATION

Discussion

AS9100 requires that a controlled system be implemented to capture communications between the organization and the customer. This reflects the contract, product, and change requirements in sections 7.2.1 and 7.2.2. Section 7.2.3 requires that these systems be designed to communicate with the customer. This includes customer feedback (a management review subject), customer complaints (see section 8.5.2 [a]), and customer satisfaction (section 8.2.1).

Customer feedback and/or complaints are good performance indicators for the overall effectiveness of the AQMS. It's important to ensure that the organization's customer communication promotes a process by which customers can provide feedback or complaints, that data are collected as part of this process, and that the data are analyzed effectively.

Outputs from this process should provide information for management review and other AQMS processes as well as enhance overall customer satisfaction and drive continual improvement.

Findings

Findings are routinely issued against this section regarding a failure to show communication of customer complaints. Also, many times there is no system to show compliance with the customer satisfaction requirement, and so the auditor will raise the finding in this section in lieu of section 8.2.1 if it appears to be due to a lack of communication.

Best Practices

If e-mail or fax communications are used for contracting purposes, be sure that the paperwork is included in the contract file or that the file has reference to them. Log sheets are useful for communications that are received indirectly (e.g., communications from the receptionist, secretary, or other nontechnical employee) since they show when the communication occurred and from whom, as well as other relevant information. The important thing in this section is the *method*, not the *message*.

NOTES

Chapter 7
Design and Development Processes

Aerospace auditors are undergoing more scrutiny because of the Industry-Controlled Other Party (ICOP) scheme. The design and development sections are routinely excluded by AS91xx registrants that have no design authority. An auditor who wishes to renew his or her certification must include at least one audit that had a design and development assessment. This has been a challenge for many auditors, as design and development aren't usually part of an AS9100 audit. This is because section 7.3 in AS9110 pertains specifically to organizations responsible for the design of modifications. (This book excludes the discussion of a very limited section 7.3 application to AS9110. It does not apply at all to AS9120.)

Subclause 7.3.1 states:

> The different design and development tasks to be carried out shall be defined according to specified safety or functional objectives of the product in accordance with customer and/or authority requirements.

Under ICOP, this is an enhanced area of scrutiny. The common thought is that the design controls for aerospace products must meet the same requirements as the end product. Before an audit, the registrar must determine how the organization handles its design and development activities. If the customer controls this process and the organization has no control over its designs, then section 7.3 can be properly excluded.

One important consideration is that AS9100 clause 7.3 refers only to design and development of products and services, not processes. In some organizations it can be beneficial to apply this methodology to processes, but it's not required.

Figure 7.1 depicts the overall design and development process, from customer inquiry to post-delivery activities, as well as references to the applicable AS9100 design and development subsection requirement. Remember that ISO 9001:2008 stresses post-delivery activities, which are an integral component of AS9100C. A turtle diagram is helpful to illustrate inputs, outputs, and performance goals for the auditor. Appendix F illustrates a sample turtle diagram.

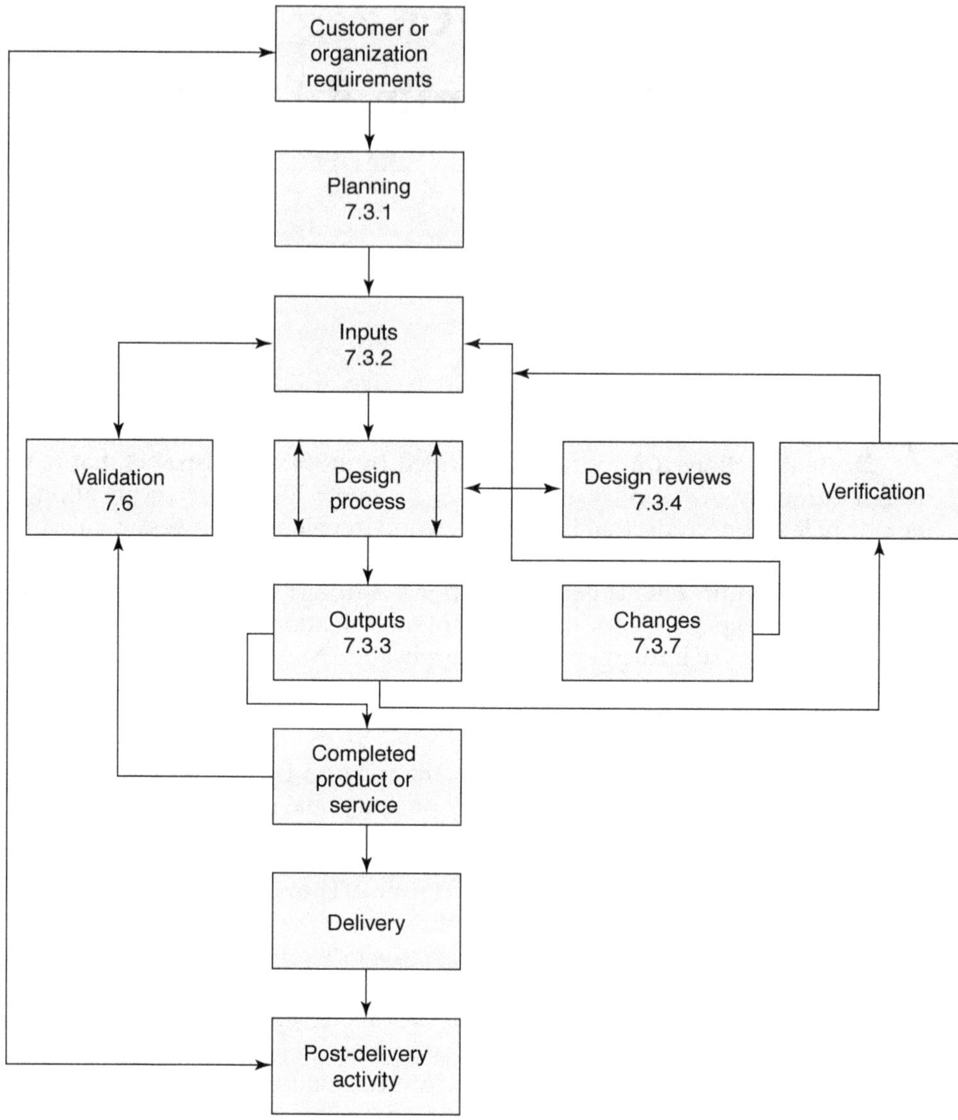

Figure 7.1 Phases of AS91xx design and development.

SECTION 7.3.1: DESIGN AND DEVELOPMENT PLANNING

Discussion

There are significant additions in AS9100's requirements for design and development. It's important to understand that the regulatory requirements thrust on the original equipment manufacturers (OEMs) for production of aircraft, engines, and accessories are complemented in section 7.3.

Section 7.3.1 sets the requirements for design planning. They are described as follows:

- What is the overall flow of the design planning process?
- Has configuration management been established as part of the design process?
- Who is responsible for design planning, and are the responsibilities and authorities defined for these individuals?
- Will any constituent part of the design be outsourced?
- Is there a planned and defined interface between various groups identified and managed, including the customer and any required regulatory personnel?
- Does the planning include the required verification, validation, and review points?
- If this is a complex design, is there a structuring of the design plan into significant elements?
  - —Is there an analysis of the tasks and the necessary resources for design and development? Does it consider an identified responsible person, design content, input data, planning constraints, and performance conditions?
  - —Is the analysis of the input data specific to each element reviewed to ensure consistency with requirements?
  - —Are there records to reflect the analysis and review?
- Does the plan identify the main milestones and timelines?
- Is the plan updated and communicated to all relevant groups as necessary?

The overall design and development plan should cover all this in a procedure or set of procedures. Remember that section 4.2.1 requires the organization to prepare documents needed to ensure the effective planning, operation, and control of its processes.

Findings

Findings related to section 7.3.1 usually concern the documentation trail for the design process. The documentation requirements issued by the organization are the usual culprit. The design process may require signatures from the parties concerned with the planning. When they're incomplete, a finding is issued. A failure to set the tasks required also raises findings. A source of major findings is a lack of required regulatory or customer approvals at the planning stage.

Best Practices

When the process for design and development is in place, back it up with a checklist that includes all the required meetings, signatures, planning dates, and changes and identifies the design team. Those individuals who must sign should be identified in the checklist. Keep this checklist in the engineer's file. Electronic records can also have a checklist. Make sure that there are suitable controls for electronic signatures. Also make sure that the design team and any other required parties are clearly defined and that authorities and responsibilities are set.

Instead of requiring that design and development review, verification, and validation have separate records, AS9100C allows organizations to perform these reviews in a way that best suits them. This allows smaller organizations to conduct validation and verification simultaneously. These records must still be maintained.

SECTION 7.3.2: DESIGN AND DEVELOPMENT INPUTS

Discussion

If the requirements in this section seem familiar, it's because these input items were derived from section 7.2. In fact, design and development personnel must be involved at the section 7.2.1 and section 7.2.2 stages if the contract includes design and development, to ensure that the requirements are "complete, unambiguous, and not in conflict" and that the functional and performance requirements are understood prior to accepting the contract. Where previous designs are used as information, there should be a clear link between the designs and the information. The applicable regulatory requirements are also a major issue here, especially for products subject to Federal Aviation Administration (FAA) certification requirements. Supplemental type certificate (STC) and Parts Manufacturer Approval (PMA) information should be readily available for the registrar. Designs under these two concepts will have a multitude of documentation from installation manuals, maintenance procedures (instructions for continued airworthiness [ICAW]), master drawing lists, PMA listings, and STC approvals.

Findings

Customer complaints regarding product design may lead to findings in this section if the complaints are not used as input items. Failure to provide or document regulatory-required inputs is a common finding. Some cases of ambiguity between the customer requirements in a statement of work or similar contract term and the design have been raised as findings. Generally, though, a lack of objective evidence causes these findings. There is so much oversight by the regulatory authorities and the customer that it appears that the principles are applied but the paperwork is missed.

Best Practices

Make sure that the design and development personnel's involvement in the contract review is reflected in the design inputs. The checklist described in the best practices for compliance with section 7.3.1 is a great tool for this purpose. An auditor may ask to see the relationship between the section 7.2 process and the section 7.3.2 process. You need to ensure that the required documentation, as well as the previous design information, is present. Make sure you can link the present job to the prior design.

SECTION 7.3.3: DESIGN AND DEVELOPMENT OUTPUTS

Discussion

The design and development outputs should comply with the organization's plans in order to ensure that the resulting product is consistent with the design requirements and suitable for its intended use. The standard focuses on verification for the outputs. It requires that the output be in a form that enables verification and that it be approved prior to release. It also requires the organization to specify the characteristics of its product that are essential for its safe and proper use. These might be temperature, pressure, or environmental requirements. It mandates configuration management and denotes the use of drawings, parts lists, and specifications to verify the product and information for materials, processes, type of manufacturing, and assembly necessary to ensure (prove) conformity.

Findings

Findings here usually surround the configuration management aspect. A lack of inspection criteria for manufacturing may also give rise to a finding. Though findings do occur, this area is not noted for a great number of them.

Best Practices

The use of a compliance matrix, such as the one in Figure 7.2, to ensure compliance with the requirements of section 7.3 is of paramount importance. Ensure that

Statement of work		**Response**			**Comments**
Section/page	Requirement	Section/page	Paragraph	Response	
Regulatory requirement		**Response**			**Comments**
Section/page	Requirement	Section/page	Paragraph	Response	

Figure 7.2 Compliance matrix template.

configuration management is addressed and planned for. Establish clear requirements for the design outputs to translate to the conformity documents. This is especially true under the FAA's certification rules. Make a compliance matrix to account for all customer and regulatory requirements. This will make it much easier during the verification stage.

AS9100C ties the definitions in section 3—including "critical items"—to output requirements. This extends the requirement that key characteristics include critical items and may result in additional procedural requirements in an organization's design guide and additional audit items.

SECTION 7.3.4: DESIGN AND DEVELOPMENT REVIEW

Discussion

The revision in this section addresses the requirement of progression authorization. The standard is silent on what reviews are necessary; it merely states that reviews are at suitable stages. The process should ensure that reviews occur at planned stages throughout the design process. This can be prearranged organizationally or at intervals designated at a part-by-part determination. The reviews must be performed in a systematic way involving representatives of the functions concerned with the stage being reviewed. The end product of the review should be the determination that the output demonstrates the suitability, adequacy, and effectiveness of the designed product or service and that design objectives are being achieved. The review must authorize progression to the next step, and there must be adequate records of the review.

Findings

The main reason for findings in this section concerns the records requirement. For example, an organization might have reviews but not maintain records of them. Or, there might be records available but no indication of an authorization to proceed to the next stage. Required personnel may be missing from the review. All of these point to flaws in the record-keeping process required in the design and development program.

Best Practices

There is no set requirement for how many reviews are necessary. The "suitable stages" requirement can vary from project to project, depending on complexity. A small modification of a prior completed product won't have as many stages as the original design project. It's important, therefore, to decide at the planning stage exactly what reviews are necessary—especially if your company doesn't employ a set design review scheme, such as a kickoff review, followed by a preliminary design review (PDR) and then a critical design review (CDR). If you choose to have only a CDR because of reduced requirements, then state this at the planning stage to avoid confusion later on. There must be records of these reviews, the actions that resulted from them, and the approval to proceed. Make sure that your action items from the review are accounted for and completed. This is why a formal master schedule and checklist are important in this process. The checklist should include what reviews are required, when they will be done (listed by chronological order rather than calendar date), who must attend, action items and their dispositions, and the all-important approval to proceed to the next step.

A good record of the review is essential. Figure 7.3 illustrates a generic design review form that satisfies the 7.3.4 requirements, although it would have to be customized before it could be used.

Design Review		
Project: ❑ Kickoff ❑ PDR ❑ CDR		
Component/system/subsystem:		
Project engineer:		**Date of review:**
Subject (check as applicable)		
❑ Requirements	❑ Hardware	❑ Quality
❑ Performance	❑ Configuration	❑ Safety
❑ Cost	❑ Materials	❑ Other ____________
Attendees (review board)		
❑ Quality	❑ Document control	❑ Production control
❑ Materials	❑ Engineering	❑ Customer (if required)
❑ Manufacturing	❑ Configuration	
Review board comment/recommendation		
❑ There are action items. (See attached list.)		
Design engineer's response and disposition		
Signature:		Date:
Authorization to proceed		
As a result of this review, this project can proceed to the next stage.		
Signature:		Date:
Design engineer's closeout		
Sign when all action items are complete.		
Signature:		Date:

Figure 7.3 Design review form.

SECTIONS 7.3.5, 7.3.6, 7.3.6.1, AND 7.3.6.2: DESIGN AND DEVELOPMENT VERIFICATION, VALIDATION, AND DOCUMENTATION

Discussion

These four subsections are discussed together because many organizations perform them simultaneously. Many customers elect to perform the validation themselves, but if they do not have the resources or the expertise, they may send the product to another vendor to perform the validation. Verification can be performed with alternative calculations; by comparing a new design specification with a similar proven design specification; by performing demonstrations, including prototypes, simulations, and tests; and by reviewing documents prior to issue.

There will often be an approved test plan that sets the parameters and criteria for verification of a design. The approved test plan may also be the validation of the design. The standard allows the validation to be done in stages, so there is no requirement to perform a single full validation. It also allows that there may be multiple intended uses and requires a validation on all of them.

Simple products may require simple validation: a test of a prototype and then the first article. More often than not, however, design verification and validation are more complex. Aerospace products usually require compliance with federal regulations and may have to comply with several specific performance requirements established by the customer, end user, or regulatory authorities. In such a situation, design validation can only be completed by obtaining information about the performance of the products or components (preferably formal test results) from such design organizations or by users of the products or components. In the case of FAA approvals for testing, various forms of approval are necessary for aircraft and components.

Some testing, especially in the case of modifications, requires extensive flight testing. Approval from the designated engineering representative (DER) and the Aircraft Certification Office (ACO) is necessary at all points during these tests, and the resulting data and analysis are strictly controlled.

Findings

There are more than just a few findings in the area of verification and validation. Most findings involve a failure to maintain a record of the verification. Auditors will evaluate projects to ensure that the customer requirements match the verified or validated results. Unsigned, unreviewed, or unapproved test plans also create a few findings. It's mainly an administrative error in this area, but major findings can occur when customer requirements aren't validated. Findings are also raised where paperwork required by the FAA/European Aviation Safety Agency (EASA) is deficient or missing.

Best Practices

The best practice for compliance with these sections is the checklist you used in checking off the requirements in sections 7.3.1–7.3.6. Make sure that you have defined in the planning stage what verification and validation actions are required. The customer cannot waive any of these requirements to your AS9100 registration. A well-defined design procedure will assist your navigation through these often-troubled waters. As part of each project, a design notebook should be created and maintained. This will provide a central repository for all of the records required by section 7.3. The notebook should contain at least the following:

- A checklist to cover all steps in planning and execution.
- Authorization to begin the project.
- Project plan with timeline showing key events.
- Definition of organizational and technical interfaces, including the responsibility and authority for all tasks.
- Design input (product specifications, regulatory requirements [or reference to them], customer requirements).
- Design output (drawings, documents, bills of material, ICAW, illustrated parts catalogs, or reference to the location where these files are kept).
- Evidence of design reviews and approvals. Unless it's a very simple project, you should have, at a minimum, evidence of a kickoff meeting, a PDR to approve the specifications, and a CDR to confirm that the product meets the specifications.
- Evidence of design verification and validation. This should include the approved test plan and test results showing performance testing for each manner in which the customer will use the product.
- History of changes to the design and specifications.
- Critical communications.

SECTION 7.3.7: CONTROL OF DESIGN AND DEVELOPMENT CHANGES

Discussion

This section is more important than many people think. All changes made during the design and development program or after its completion must be documented. This includes regulatory authorities. Most organizations use engineering change notices (ECNs) and engineering change orders (ECOs), which are the most common methods used to satisfy the control, document, and record requirements of section 7.3.7. The same elements of verification and validation apply here as well as to the main design program. If the change is complex and alters the product's overall form, fit, or function, a full design and development project may be warranted. An example of this would be changing the engine make or model in an existing aircraft design. Records are required of the review of the changes and any actions made. Finally, note that there is also a requirement for approval before implementation.

Findings

Findings are common in this area because changes are often made but not documented with review and approval. Some organizations use this process to control documentation, that is, issuing an ECO for a change in process documentation (e.g., quality manual). Findings will arise because the documentation changes are not approved in accordance with the ECO procedure. Also, not having the appropriate number of signatures, or not having the specified individuals' signatures, on the process is a source of many findings.

Best Practices

The changes in AS9100C section 7.3.7 reflect the updated standard's increased focus on configuration management. This section requires that all design and development changes be controlled by the configuration management process. In most organizations, however, this has been common practice.

Try to avoid mixing the approvals for the engineering department with the system and procedures of the aerospace quality management system (AQMS) documents. Make sure that your record for the change—whether it's an ECO, ECN, engineering change request (ECR), or whatever other terminology you choose—covers the review necessary, including verification and validation. An example of an ECR record is shown in Figure 7.4.

Tailor these records to your organization's needs. If you don't need all of the blocks shown, place "N/A" or some similar note in the block to show that it wasn't an oversight.

Make sure that these records are integrated into your configuration management plan and document control procedures. Lay out the entire change process in a flowchart and ensure that all of its elements align with the standard and any customer/regulatory requirements. STCs are a form of design change and strictly controlled within the federal aviation regulations. Make sure that your procedures account for these requirements.

<table>
<tr><td colspan="3">Engineering Change Request</td><td>ECR no.</td></tr>
<tr><td>Document/item no.</td><td>Current revision</td><td>Document title</td><td>ECR class</td></tr>
<tr><td>Requester name</td><td>Request date</td><td>Requesting organization/department</td><td>ECR priority</td></tr>
<tr><td colspan="4">Need for change Primary justification classification [D] Interface
1. How and when was need recognized (e.g., test results, field reports, engineering review, production problems)?
______
______
2. What is the effect of not making the change (e.g., safety hazard, mission failure, schedule slippage)?
______
______
3. How will change improve the product/system (e.g., increased reliability, reduced weight, decreased cost, improved performance)?
______
______</td></tr>
<tr><td colspan="4">Documents/data to change (check all that apply)
❑ Print ❑ Bill of material ❑ Router ❑ Technique master ❑ Item master</td></tr>
<tr><td colspan="4">Description of change (what? why? how?)
______
______
______
______
______</td></tr>
<tr><td colspan="4">Approvals and disposition</td></tr>
<tr><td>Lead engineer (name)</td><td>❑ Approve
❑ Reject</td><td>Signature</td><td>Date</td></tr>
<tr><td>Engineering manager (name)</td><td>❑ Approve
❑ Reject</td><td>Signature</td><td>Date</td></tr>
<tr><td>Operations (name)</td><td>❑ Approve
❑ Reject</td><td>Signature</td><td>Date</td></tr>
<tr><td>Customer approval</td><td>❑ Approve
❑ Reject</td><td>Signature</td><td>Date</td></tr>
<tr><td>Disposition
❑ Implement
❑ Hold for next document change
❑ Reject
Date ______ by ______</td><td colspan="3">Reason for disposition/remarks/instructions
______
______
______
______</td></tr>
</table>

Figure 7.4 Engineering change request.

NOTES

Chapter 8
Purchasing

The purchasing provisions of section 7.4 are divided into three distinct subsections: 7.4.1, 7.4.2, and 7.4.3. They are separate because each subsection's operations relate to different functions within the organization. Section 7.4.1 addresses the control of suppliers, which is usually handled by the quality department. Section 7.4.2 concerns purchase order information, which is usually handled by the purchasing and materials department. Finally, section 7.4.3 relates to the verification of purchased products, which is usually handled by the quality department through receiving inspection.

The changes to this section of the standard address the critical processes that most concern original equipment manufacturers (OEMs). If the supply chain cannot provide good product, then the end-user OEM can't rely on the integrity of the final product. This is one of the reasons that the Industry-Controlled Other Party (ICOP) scheme was developed and that there is such strong emphasis on publications concerning the supply chain, most notably the Supply Chain Management Handbook (found on the OASIS website). The amount of oversight of aviation suppliers is increasing, and even the industry's registrars are under scrutiny.

AS9104/1 requires that the purchasing process be audited annually. Make sure that the purchasing process is thoroughly internally audited to prevent recurring findings during third-party audits, as recurrent findings are automatic major findings under section 8.5.2.

SECTION 7.4.1: PURCHASING PROCESS

Discussion

This subsection concerns the selection, control, and disapproval of suppliers. It represents a quantum leap over ISO 9001's requirements and begins with the basic ISO 9001 provisions:

> Criteria for selection, evaluation, and re-evaluation shall be established. Records of the results of evaluations and any necessary actions arising from the evaluation shall be maintained.

This is one area in which a document will be necessary to establish the criteria. The term "reevaluation" is subject to at least two definitions:

- To periodically review all suppliers
- To review suppliers with problems or disqualifications

AS9100's requirement to periodically review supplier performance bolsters the interpretation that there will be a periodic review of all suppliers. It would be contrary to the tenets of the AS9100 scheme and FAR 21 requirements if suppliers were never reevaluated.

Procurement is a critical process for the aerospace industry. There should be some proactive internal process to ensure that potential suppliers can provide a product that meets the design specification at the determined cost. The organization should ensure that the specified purchase requirements are correct prior to their communication to the supplier and that the supplier is capable—practically and legally—of providing the product. The control of suppliers is essential.

The difficulty for most new organizations is that this section can be read to be very labor-intensive, much more than is reasonably required. There is an "escape clause" within this language:

> The type and extent of control applied to the supplier and the purchased product shall be dependent upon the effect of the purchased product on subsequent product realization or the final product.

This doesn't mean that you have to control evenly across the supplier spectrum. The toilet paper your organization buys at Costco won't be rated very high on control (unless toilet paper is an important constituent part of your product, of course). Auditor opinions vary greatly in this area, and you should be able to demonstrate your process and explain why you don't require an annual on-site audit of Costco to be able to buy toilet paper there. Most auditors will accept your process if your requirements are logical. The important concept is that the more critical the vendor-supplied product or service is to your final product or service, the more control you need to influence on the vendor.

An approved supplier list (ASL) is required, and any supplier that affects your product or service should be listed. This list must show each vendor's scope of approval to demonstrate which goods or services are available to be purchased from each vendor and the status of each vendor (e.g., approved, on probation, disqualified). One vendor may be approved for everything that it sells and another for only a singular part, material, or process.

The organization must rate the performance of suppliers. There are many ways to do this. The standard is silent on its frequency. The other provisions in this subsection require the procedure to show that the function that has approval authority also has disapproval authority. It also requires that approved suppliers use approved subsuppliers, where applicable.

AS91xx also requires the organization to perform a risk analysis for suppliers.

AS9120 adds one more requirement: The organization must implement controls to prevent the purchase of counterfeit and suspected unapproved parts.

Findings

There are many findings in this area during the formative stages of registration. Fortunately, they tend to diminish once the organization is certified and progressing with continual improvement actions. Most of these early findings concern the use of suppliers that are not on the ASL or incomplete records of supplier reviews and any actions taken. Vendor listings without scopes of approvals also warrant findings. Procedures such as the criteria for selection, evaluation, and reevaluation are sometimes missing and result in minor findings. As time from the stage 2 audit passes, the reevaluations are missed and there are cases where the performance measuring is missed. An important, but often misunderstood, issue concerns this subsection and section 8.4:

> The organization shall determine, collect, and analyze appropriate data to demonstrate the suitability and effectiveness of the quality management system and to evaluate where continual improvement of the effectiveness of the quality management system can be made. This shall include data generated as a result of monitoring and measurement and from other relevant sources. The analysis of data shall provide information relating to . . . d) suppliers.

The registrar will expect that the data generated from your performance review of suppliers will be reviewed at the management review or another similar review to make use of them. When there is no analysis, findings are issued. If the auditor suspects an unapproved, undefined parts program, this can lead to a finding.

AS9110 requires a little extra here: The organization shall ensure that suppliers hold the required approvals and certificates. Additionally, the organization's purchasing process shall satisfy applicable authority requirements pertaining to the use of noncertificated suppliers. Failure to comply with this requirement results in findings.

AS9120 adds the counterfeit and Suspected Unapproved Parts (SUP) requirements. Failure to implement or follow these processes will result in a finding.

Best Practices

There are many ways to handle the supplier issue. Most organizations are moving to electronic databases and programs to handle their supplier oversight, but they often leave procedure and performance issues out of these programs. Keep in mind the "escape clause" in section 7.4.1 regarding this issue. Above all, remember that despite an organization's use of and reliance on a supplier, the organization remains responsible for any product it provides to a customer.

Most large aerospace repair companies have addressed this issue successfully for many years. They "tier" their suppliers on the basis of defined criteria. A good policy is to look at the types of products and services you procure and their effect on the final output to your customer. Figure 8.1 shows an example of a tiered system in a repair station that operates under the requirements of AS91xx.

The example shown in Figure 8.1 is one method for addressing different supplier classifications. Note that special process providers and tool calibration companies would be critical suppliers.

Accordingly, the performance of tier-one suppliers should be measured, but it's not necessary to measure the performance of tier-two suppliers. A good practice is to rate the performance of their quality and on-time delivery. The frequency of these performance calculations should match the frequency of the organization's management review. The procedures should dictate what to do if the supplier falls below an acceptable level (e.g., removal from list, increased scrutiny).

In the case where the vast majority of purchases come from so many suppliers that it's burdensome to do a performance review on them, it may be easier to do the performance review on the suppliers that create the bulk of your purchases. For example, if your organization has 40 suppliers but 85% of the product is purchased from 5 of them, it may be better to limit your performance reviews to the top 5 suppliers. As long as you can substantiate this approach, most auditors won't question it.

Unless you use customer-directed suppliers for other purposes, they can't be considered "your" supplier and compose a third tier. You should list them on the ASL, but your customer has the responsibility to oversee them. If you can't take them off the list in the event that their on-time delivery rates fall, ensure that you have the data to show that they're customer-driven sources.

AS9100C allows organizations to use accredited third-party registrations as part of their evaluation/reevaluation processes. This could be an ISO, aerospace, Nadcap, or NATO alliance document. If you consider this kind of document as part of the process, you must have a current certificate available. There is nothing worse than an auditor finding out that a vendor has lost or given up certification, unbeknownst to the organization.

Our approved tier-one list comprises critical vendors and service companies that appear on the function list that we can use in manufacturing. Our approved tier-two list comprises suppliers of parts that our organization can use in manufacturing.

6.1 Our organization will not use any outside tier-one vendor until that vendor has been approved in accordance with the following procedure

- 6.1.1 The chief inspector will:
  - 6.1.1.1 Use checklist aid WP 15-3 in qualifying any vendor.
  - 6.1.1.2 Assemble the audit team, to include a team leader and at least one other auditor, but not more than three auditors. The auditors shall not be from the same area within our organization, which means, for example, that the team will not be made up of three members from the inspection department. This will ensure diversity in experience in evaluation of contractors.

6.2 Procedures for the surveillance of the tier-one contractor

- 6.2.1 Each procurement from such a vendor will be completely inspected by our organization's inspection department upon return to our organization in accordance with OP 6.0 as if it were a customer-incoming part along with all material and/or process certifications necessary for traceability of the article. The details of the inspection will be placed on the traveler in the appropriate location, and the certifications will be attached to the traveler. If an article does not meet specifications, a discrepancy report will be generated by the inspector finding such nonconformance and routed to the chief inspector.

Figure 8.1 AS9100-compliant tiered rating system example.

- 6.2.2 The chief inspector will evaluate the findings and make a disposition to the quality control individual serving as the point of contact for the vendor. The vendor will respond to the discrepancy report within fifteen days. If the vendor fails to respond, the vendor will be removed from the approved contract maintenance list until the discrepancy report can be cleared.
- 6.2.3 The chief inspector, or the person he or she designates, will visit the vendor at irregular times to verify that the information on the initial form WP 15-3 is still applicable to the vendor. Each noncertificated vendor will be visited at least once per twenty-four-month period, but may be visited more if the number of discrepancy reports indicates a systemic problem.
- 6.2.4 The chief inspector, or the person he or she designates, will send our organization form WP 15-5, along with a cover letter similar to the one contained in WP 15-5, to the vendor as a self-audit to be conducted by the vendor at least once biannually on the year after the surveillance set forth in 6.2.3.
- 6.2.5 This type of vendor must allow reasonable inspections by the customer, the regulatory agencies, and our organization to observe the vendor's work on the article.

6.3 Procedures for the surveillance of the tier-two contractor

- 6.3.1 All parts purchased and received from the approved tier-two vendor shall be checked by the material handler in the purchasing department for shipping damage and sufficient documentation to ensure traceability of the parts to an FAA-approved source.
- 6.3.2 The material handler will forward the traceability paperwork to the inspection department for review and approval before putting the parts into stock or service.
- 6.3.3 The director of quality control will ensure the following:
  - 6.3.3.1 Each article received from such a vendor will be completely inspected by the material handler upon receipt at our organization as if it were an incoming part, along with all material and/or process certifications necessary for traceability of the article. If an article does not meet specifications, a discrepancy report will be generated by the inspector finding such nonconformance and routed to the chief inspector.
  - 6.3.3.2 As long as a tier-two vendor performs properly with respect to condition and traceability, it may be maintained on the tier-two list. If a tier-two vendor forwards traceability documentation that is suspect or deficient, the director of quality shall be notified. If the problem appears to be chronic or systemic, the director of quality may remove the vendor from the tier-two list.

6.4 Approved list of tier-one and tier-two contractor and vendor procedures to properly maintain the list

- 6.4.1 The approved vendors list is a separate publication maintained in the quality office with a copy of the tier-two vendors list in the purchasing office, and is revised as necessary to comply with the procedures set forth above by the director of quality control.
- 6.4.2 Technical training on contracted functions on tier-one vendor items for receiving inspection personnel and procedures for receiving inspections that provide enough technical detail to determine acceptability of an article.
  - 6.4.2.1 The receiving or processing inspector will be trained and knowledgeable on all requirements for any inspected item in accordance with OP 5.0.
- 6.4.3 If the approved tier-one vendor requires a third-party certification, the director of quality control will ensure the following:
  - 6.4.3.1 That the contracted vendor is properly certificated, not just compliant, to provide the required material or service. Items received from a certificated facility must be properly processed through the receiving inspection procedures set forth in OP 6.0, as our organization is ultimately responsible for all the work performed on parts used in the manufacture of the article being approved for return to service. A copy of the certification will be obtained and made part of the vendor file.

Figure 8.1 AS9100-compliant tiered rating system example. *(Continued)*

This is an important subsection, and care should be taken to ensure that you've accounted for each of its requirements in your processes. Remember that it is audited annually. Additionally, be sure to have a good SUP or counterfeit program in place, with training given.

SECTION 7.4.2: PURCHASING INFORMATION

Discussion

A significant number of requirements have been added to this section of the standard. Many of them have the "where appropriate" qualifier. If they are not a requirement of your process, customer, or regulatory authorities, it is not necessary to register them to AS91xx.

The best example of this is the requirement for test specimens. If your process doesn't require a test specimen, and your customer doesn't require it, it doesn't apply and need not be addressed by the registrar. To be on the safe side, some of the requirements (such as right of access) should be discussed. This is usually a function of the purchasing department, so the AS9100 team should confer with the purchasing department on its need for purchasing information. AS91xx section 1.1 states the following:

> This International Standard specifies requirements for a quality management system where an organization . . . a) needs to demonstrate its ability to consistently provide product that meets customer and applicable regulatory requirements.

This means that objective evidence will be necessary to show compliance. The nexus of "where appropriate" addresses the organization's actual processes.

The organization should ensure that the specification quoted in a purchase order—whether as a product or a service—is the same as the specification contained in the design (or the specification received from the customer). It should discuss, internally and with its potential suppliers, the design specification of procured items during the design process or prior to order placement. There should be some form of approval of the specification before the final specification/order is confirmed to the supplier. The registrar will evaluate all of this during an audit. Finally, any statutory or regulatory requirements should be accounted for in the purchase documents.

Findings

Most findings in this area concern missing information from purchase order documents. For example, the customer contract may require the use of suppliers that are certified under a quality management system (QMS) such as AS9100, ISO 9001:2008, or ISO 17025, and this designation isn't included in purchase documents. Another common finding is for a customer specification not to align with the corresponding specification generated by the organization—for example, a purchasing department requests an old specification for a heat-treating process but the supplier performs the process to a new specification. Specified release documents may not be available, such as FAA form 8130.3, and this could also be a finding. If an organization knows that a supplier will be subcontracting a process, the purchase information should show the flow-down of this requirement; this is missing on occasion.

Best Practices

There should be a terms and conditions page issued to suppliers that have these purchase terms included. Make sure that the quality department is aware of the need for quality-type clauses and that it has included them in the terms and conditions document or issued a separate quality clause requirement on the purchase order. Figure 8.2 shows an example of such a document.

There should be a clear channel of communication between purchasing and the quality department on these issues.

Boogaloo Bearings Supplier Terms and Conditions

A certification of compliance addressing each item or lot of material shipped against this purchase order must accompany each shipment. BBI part numbers, Mil-Spec devices, or supplier equivalent identification (see further details below) are referenced as part of the certification format.

Mil-Spec fasteners used shall be of the latest revision of the applicable specification when received. Supplier or sub-tier supplier certifications for specific processes or tests are acceptable provided they reference the exact requirements (i.e., drawing, specification, test, or process that was performed) on their certification. Material, chemical, or physical test reports for processes performed by suppliers are kept on file at the supplier for verification.

Each primary supplier certification shall include:

- Purchase order number
- Part number and revision, as shown on the BBI drawing. For an assembly, include the assembly part number, any subcomponent part number(s) fabricated by the supplier for this assembly, and all material and processes used therein if required in the purchase order. For raw stock, a supplier part number or simple description is acceptable when traceable to the BBI part number. In this case the certificate of conformity is annotated with the applicable BBI part number and acceptance by an internal quality engineer and revision letter.
  - When a specific revision is required by the purchase order, the certificate of conformity must reflect that the item is in compliance with the requested revision. Some items, such as raw materials (i.e., bar stock), may not have a revision. Otherwise, it is accepted that the item is to the latest drawing and/or specification revision to the date of manufacture.
- Quantity shipped
- Lot number (as applicable)
- The name and address of the company issuing the certification
- The date of issue
- The name of the company's authorized representative (optional)
- The representative's title (optional)

A certification for ASTM, ASME, SAE, MS, BAC, AN, or NAS specification fastener(s), bolts, nuts, screws, studs, washers, rivets, pins, etc., is not required.

BBI-5A sampling inspection: Sampling inspection may be performed to accepted standards or to BBI-approved sampling plan.

Figure 8.2 Terms and conditions example. *(continued)*

BBI-5B 100% inspection requirement: Supplier shall perform 100% functional, electrical, and dimensional inspection, as appropriate, on finished goods to ensure integrity of the item/lot.

BBI-6 records: The supplier shall maintain records for review by BBI or its customer/regulatory agency for a minimum of seven years. Records include material certification, special processing, work orders and travelers, test reports, inspection reports, calibration records, and first articles. If required by purchase order, electronic media for nonconforming programs, test programs, or drawings may be required for archive purposes at the direction of the BBI buyer.

BBI-7 chemical and/or physical test reports: One copy (as required by specification for each lot, batch, or heat) must be attached to the packing sheet and must accompany each delivered shipment. The report must be identifiable to the product being delivered and must show that the material or parts delivered are acceptable.

BBI-8 shelf-life items: Materials or articles having characteristics subject to degradation with age shall be marked in a manner to indicate the date of manufacture and expiration date in addition to standard identification requirements. Suppliers of rubber goods are required to document the cure date of rubber molded goods and the compounding date of uncured rubber. Materials identified as requiring age-sensitive controls must have a minimum of 80% of their expected shelf life from the date of manufacture at the time of receipt at BBI. Special storage or handling conditions shall be included when applicable. Material safety data sheets must be provided where there is a danger to health from material being purchased.

BBI-9 packing/packaging: All items must be adequately packed/packaged to prevent damage, contamination, and/or deterioration.

Figure 8.2 Terms and conditions example. *(Continued)*

SECTION 7.4.3: VERIFICATION OF PURCHASED PRODUCT

Discussion

The verification activities mandated in section 7.4.3 include several clauses that pertain to examples of objective evidence of the quality of the purchased product or service, inspection of these goods or services at the vendors' premises, review of documentation, inspection upon receipt, and the delegation of verification to the vendor. AS91xx section 1.1 states the following:

> This International Standard specifies requirements for a quality management system where an organization . . . a) needs to demonstrate its ability to consistently provide product that meets customer and applicable regulatory requirements.

Objective evidence will be necessary to show compliance and the nexus of any requirement that is deemed "where appropriate" in the organization's processes. The clause also includes an additional requirement for the verification activity, that is, that the purchased good or service not be used or processed until it's been verified as conforming to specified requirements, unless it's released under positive recall.

If test reports are used as objective evidence, they must conform to the standard's "applicable specifications." AS9100B required a periodic validation of test results, but this requirement was removed from revision C. If you choose to delegate the acceptance of the goods or services to the vendor, the requirements for delegation should be defined and a register of delegations must be maintained.

As a rule, the organization should ensure that the specification quoted in a purchase order, whether as a product or a service, is the same as the specification contained in the design (or the specification received from the customer). If there are statutory or regulatory requirements, account for them in the purchase documents and the receiving and acceptance documentation and records.

Findings

Many findings in this area arise due to the lack of certification data from the vendor. Sometimes certificates of conformity are signed by nonquality individuals from the vendor, and this might rise to a finding if the organization's procedure requires a signature from the quality department. A lack of training in this area also raises findings. Any delegation of work must meet the requirements of this section, and a definition (procedure) must be available and the process maintained. This area of delegated approval causes the most findings. Delegation processes often fall into disrepair or become outdated.

Best Practices

Avoid delegating work if you can. It can result in many findings if not handled carefully. The receiving inspection department should be the main line of defense for bogus or counterfeit parts, and this procedure should be very thorough and conducted by well-trained people. One area to be aware of is the ability of receiving inspection personnel to interpret vague or complex test and validation reports. If the report doesn't indicate compliance with the material specification, how does the reviewer know it's proper for use? Ensure that certificates of conformity are signed by someone from the supplier's quality function.

NOTES

NOTES

Chapter 9
Production and Service Provision

SECTION 7.5.1: CONTROL OF PRODUCTION AND SERVICE PROVISION

Discussion

Section 7.5.1 should be considered with the requirements in section 7.1, which pertains to planning requirements. The list of controlled conditions required to be planned for varies by type of organization and its processes relating to production.

The applicability of this section to AS91xx concerns the process of physically controlling the parts and materials within the organization's overall handling of the products. The extent of control, according to those enumerated subclauses, is:

- The availability of information that describes the characteristics of the product. This could be the specifications of the part or material. Examples include AMS specifications for materials such as 316/316L Stainless Steel Sheet & Coil, AMS 5524/AMS 5507, or, in the case of a part such as a temperature gage, MIL-G-18997E.
- The availability of work instructions, as necessary. This would include the handling of materials and storage requirements.
- The use of suitable equipment. This would include special handling equipment such as the freezer requirements for prepreg materials for suitable gloves. (Note: "Prepreg" is a term for "pre-impregnated" composite fibers. These usually take the form of a weave or are unidirectional. They already contain an amount of the matrix material used to bond them together and to other components during manufacture. The prepreg materials are usually stored in cooled areas because activation is most commonly done by heat.)
- The availability and use of monitoring and measuring devices. As with the freezer requirements given in the previous point, for example, you must make sure that you have adequate calibration of the freezer requirement. (See section 7.6.)

- The implementation of monitoring and measurement. This means a system to ensure that the monitoring and measurement processes are performed, including training of personnel assigned to these tasks. The requirements here are used as examples, but the many parts and materials used in aerospace applications may require a lot of these little processes to make sure that the product is controlled while it is in the possession of an organization that's registered to AS9100.
- The implementation of release, delivery, and post-delivery activities. This requirement is an extension of ISO 9001:2008's customer satisfaction and post-delivery requirements.
- Under AS9100, the organization must be accountable for all product use during manufacture (servicing). Examples given include quantities, split orders, and nonconforming material.
- The requirement that there be evidence that all operations, including inspection, be completed as planned is straightforward and requires signatures in the appropriate places.
- In AS9110 this planning requirement is expanded to include evidence that all maintenance and inspection/verification operations have been completed as planned in accordance with technical data that have been approved, as applicable, by either the design data approval holder or an approved design organization or are "acceptable" to the authority.
- The organization should have a foreign object detection and removal program.
- The requirement for the control and monitoring of utilities may seem overly simplistic at first glance. However, this is a common source of findings because some overhaul manuals (OHMs) or component maintenance manuals (CMMs) require dry air or, in the case of some sand- or seed-blasting processes, air pressure regulated to below a certain maximum value. This means that the pressure gage and/or pressure regulator must be contained on the organization's tool calibration recall list under section 7.6.
- The criteria for workmanship requirement shouldn't create much additional work for AS9100 registrants. Depending on the process, workmanship standards can be for the specific tasks undertaken as part of the manufacturing process. Standard practices, such as safety wires, torque values, or hardware identification, can be referenced in documents such as AC 43.13.1a. Some original equipment manufacturers (OEMs), such as Boeing, include standard practices in their maintenance manual systems under ATA (Air Transport Association of America) code section 20-00-00. Other companies draft their own workmanship standards.

AS9110 has three additional requirements:

- Maintain a list of approved maintenance/repair process capabilities and/or current ratings

- Ensure that maintenance operations do not adversely affect areas outside the scope of the planned maintenance
- Ensure that the equipment, tools, and materials are those recommended by the manufacturer of the article or are at least equivalent to those recommended by the manufacturer and are acceptable to the customer and/or authority

Findings

Calibration issues make up the largest number of findings in this area. If environmental, shelf life, or storage requirements aren't fully documented, a finding, usually major in nature, will be issued. The reason for a major finding is that the nonconformity affects the product status, including already-delivered product.

With regard to work instructions, most findings include the availability of work instructions: the lack of work instructions; the wrong work instructions; or work instructions for a different aircraft, component, or part than that being worked on. Findings are also issued in the area of paragraph (h), which requires evidence that all operations, including inspection, have been completed. This usually occurs when blanks in checklists or assembly forms are unsigned. Due to the nature of the operations, other findings related to this subsection are rare.

In AS9110 the tool equivalency requirement is the most fertile ground for findings, most of which are major. As obsolescence takes place, the availability of original specified tooling decreases. Companies that produced specified fixtures, tools, and equipment have long since gone the way of the dodo bird, and it is nearly impossible to find the exact specified tooling. This does not absolve the organization from showing that the tool that is used is equivalent to the original specified tooling.

Best Practices

There are many ways to comply with section 7.5.1. If an organization has had an aerospace presence for more than 10 years, little effort will be needed to comply with this section. This section's requirements are at the heart of any manufacturing organization. Consider how material movement is affected by its requirements. Although it's no longer used by the military, MIL-STD-980 has excellent guidance about foreign object and debris (FOD) programs. Ensure that your planning includes control plans and addresses manufacturing process controls, and that the work instructions for employees having process operations responsibility are documented, accessible at the workstation, and derived from sources such as the organization's control plan and quality plan. These plans should complement rather than contradict each other.

The tool equivalency issue is not easily resolved. It will take something along the lines of an engineer to determine that the tool in use is equivalent in all respects to the tool called out in the CMM. Prepare a procedure if your company repairs "legacy" products. Use a comparison sheet that shows what the old tool was and what the current tool is. This should include the evaluator's identification. This is not just new to AS9110; FAR 145.109 (c) has the exact same requirement for the Federal Aviation Administration (FAA) repair stations. Refer to Appendix A for the FAA's take on the process.

Ensure that you have provisions for verification of job setups according to work instructions and that the intervals for verification occur at key points, such as at the initial run of a job, during material changeover, at a job change, or during in-process engineering changes. Note whether statistical methods of verification are used.

Finally, make sure that your statutory, regulatory, and customer requirements align with those of the standard and that this alignment doesn't require a double-entry system.

SECTION 7.5.1.1: PRODUCTION PROCESS VERIFICATION (AS9100)

Discussion

Compliance with this section ensures that an organization's first-article/first-piece process performs in accordance with its documentation, which includes approved technical data.

There has always been some question as to whether the first-article inspection process belonged in AS9100's section 8. It's not a monitoring or measuring process, but a process used to ensure that a product is manufactured under controlled conditions. Because revision B had the first-article inspection requirement in section 8, the organization couldn't exclude the requirements of section 8.2.4.2. (Remember that only clauses in section 7 are subject to exclusion.)

This often becomes a minor arguing point between auditors and auditees. The placement of first-article inspection requirements under section 7.5.1.1 corrects any misconceptions about the excludability of first-article inspection requirements. In practice, this change affects only the excludability of first-article requirements, not their subject matter.

Findings

Findings or nonconformities related to this section are extremely rare. Repair stations usually have a good handle on process verification because of heavy oversight from their customers and regulatory authorities. Manufacturing organizations sometimes have problems with configuration management issues that trickle down to the production process verification area. These findings include outdated, obsolete, or incorrect technical data, drawings, and specifications.

Best Practices

Ensure that any maintenance process you use has been approved with "approved data" or "acceptable data." It's always wise to get customer approval to use any process outside of the basic maintenance manual. In some cases, the customer may require the use of a customer-specific or online first-article inspection report form. Make sure that your process accounts for the use of such forms. This doesn't negate the requirement for a first-article inspection process on products that aren't defined by a particular customer-driven program. Consider adopting and implementing AS9102 and its attendant forms, one of which is shown in Figure 9.1.

SAE AS9102 Revision A			
Form 1: Part number accountability			
1. Part number	2. Part name	3. Serial number	4. First-article inspection (FAI) report number
5. Part revision level	6. Drawing number	7. Drawing revision level	8. Additional changes
9. Manufacturing process reference	10. Organization name	11. Supplier code	12. P.O. number
13. Detail FAI ❑ Assembly FAI ❑	14. Full FAI ❑ Partial FAI ❑ Reason for partial FAI	Baseline part number including revision level	
a) If above part number is a detail part, go to field 19. b) If above part number is an assembly, go to the "INDEX" section below.			
INDEX of part numbers or subassembly numbers required to make the assembly noted			
15. Part number	16. Part name	17. Serial part number	18. FAI report number
1) A signature indicates that all characteristics are accounted for, are moot drawing requirements, or are properly documented for disposition. 2) Indicate whether the FAI is complete per section 5.4: ❑ FAI complete ❑ FAI not complete			
19. Signature		20. Date	
21. Reviewed by		22. Date	
23. Customer approval		24. Date	

Figure 9.1 SAE AS9102 Revision A.

Experienced aerospace manufacturers shouldn't require additional effort to comply with section 7.5.1.1. Organizations that produce aircraft parts under FAR 21, with an accepted fabrication manual, must ensure that their aerospace quality management systems (AQMSs) don't conflict with their fabrication manuals. The interface with document control and the organization's configuration management program should ensure that it uses only approved data. Incorporate the preventive maintenance program into section 7.5.1.1's requirement for a list of specific and nonspecific tools and nonconforming programs to avoid errors and duplication. Simplify, simplify, simplify!

SECTION 7.5.1.1: MAINTENANCE DOCUMENTATION (AS9110)

Discussion

Compliance with this section will ensure that your maintenance operations perform work in accordance with the established documentation, which includes approved technical data, including a CMM, an OHM, airworthiness directives, service bulletins, alternate means of compliance (AMOC), service instructions, maintenance manuals, or any other "approved data." This should not result in any additional work above that required by the authority having jurisdiction over the repair station operations.

Findings

It would be extremely rare for findings or nonconformities to be generated in this subsection. Any finding would probably be a violation of the authority regulations. Most repair stations have a good grasp of the requirements for this area. Findings that do exist usually concern out-of-date, obsolete, or incorrect technical data, and they usually constitute major findings.

Best Practices

As previously discussed, little additional effort will be needed in order to comply with this section of the standard. If the repair station operation is fully compliant with the authority regulations, no additional effort will be needed. One thing to be aware of is to make sure that your regulatory system does not require a double entry into your aerospace system. Some operators choose to incorporate the quality control system required by FAR 145 (Transport Canada Subpart 73) and the AQMS required by AS9110. The only guidance offered here is that if you choose to make the regulatory manual your AQMS manual, you will need to submit changes to the FAA office or the Transport Canada office for acceptance each time you modify your AS9110 system. In any case, make sure there is a correlation between the standard and your authority regulatory procedures and manuals and that it does not require a double-entry system, which could lead to findings.

SECTION 7.5.1.2: CONTROL OF PRODUCTION PROCESS CHANGES (AS9100)

Discussion

This section requires that any position authorized to approve production process changes be identified. The requirement includes a provision that any change affecting processes, equipment, tools, and programs be the subject of procedure. Further, the results of changes must be assessed.

Findings

Findings related to this section are common. Generally, a finding involves a change to a production process by a person not identified in a procedure, process, or work instruction. Sometimes the lack of "procedures . . . to control . . . implementation" of process changes gives rise to a finding.

The final requirement in section 7.5.1.2 requires that changes to production processes be assessed to confirm that their desired effect has been achieved without any adverse effects on the product. However, processes are often changed without a subsequent assessment, resulting in a finding.

Best Practices

An organization's documented procedures must include the identity, by position, of the persons authorized to change production processes and production planning. A matrix, such as the one illustrated in Figure 9.2, is helpful for showing the different levels of approval and the individuals authorized to approve the various levels of production changes.

Minor production changes that can be verified at final inspection probably don't need to be formally assessed to determine whether there are adverse effects to the product. Substantial changes, however, should have some type of engineering approval such as an engineering change order (ECO) or engineering change request (ECR), both of which evaluate potential changes to the product. Make an approval matrix that shows which signatures are required for each type of document.

Document type	Engineering manager	Program manager	Program engineer	Quality manager	Quality engineer	Production engineer	Production manager	Materials manager	Document control	Customer
Quality procedure				1	2				AR	
Work instruction				1	2	1	1		AR	
Drawing	2	1		1	2				AR	WR
Bill of material		1	2		1		1	1	AR	
ECR	2	1	1	1	2	1	2	1	AR	
ECO	2	1	1	2	1	1		1	AR	WR
1: Primary authority 2: Secondary authority						AR: Always required WR: When required				

Figure 9.2 Approvals.

SECTION 7.5.1.2: CONTROL OF MAINTENANCE PROCESS CHANGES (AS9110)

Discussion

Unless your repair station is a designated alteration station, or your organization has a designated engineering representative (DER) or designated airworthiness representative (DAR) on staff, this section does not apply to you. Very few repair stations have the capability or approval to change maintenance processes without advance regulatory approval. The only exception to this interpretation is the rare case of an AMOC. There are times when compliance with an airworthiness directive can be effected with an alternate means and the methodology for obtaining an AMOC is found at the bottom of each airworthiness directive.

Findings

Because this is a limited applicability subsection, very few, if any, findings are raised in this area.

Best Practices

Make sure that in the event of a process change, your organization acquires customer and/or authority approval. If your company is seeking an AMOC, make sure it is fully approved before undergoing compliance with the alternate means.

SECTION 7.5.1.3: CONTROL OF PRODUCTION EQUIPMENT, TOOLS, AND SOFTWARE PROGRAMS (AS9100)

Discussion

The most important part of this subsection—and potentially the most damaging—is found in its first paragraph. It requires that all production equipment, tools, and programs be validated prior to use. If the drawing, traveler, specification, or process requires specific equipment, tools, or programs, they must be validated prior to use. Further, this equipment must be maintained and inspected periodically according to documented procedures. Historically, these documented procedures included manufacturers' recommendations and operating instructions for production equipment, special tools, and software programs, including computer numerical control (CNC) software.

This section also requires the organization to establish storage requirements that include periodic checks of equipment and tooling in storage. Auditors will seek to ensure that any equipment put in storage is periodically inspected according to an established procedure. The registrar will want to verify that CNC equipment and its programs are validated prior to use.

Findings

Findings are issued under this section when production equipment, tools, and software programs cannot be shown to be validated prior to use. If it can be shown that this failure affected the quality of the product, a major finding may be issued. This would also trigger recall or notification to the organization's customers. Minor findings are routinely written, usually for equipment placed in storage with no preservation check or any procedure directing the implementation of storage requirements, including periodic preservation/condition checks.

Findings are also issued when damaged monitoring and measuring equipment is found in the production area. This section may include references to section 7.6, Control of monitoring and measuring equipment. Periodic inspection often includes calibration.

Best Practices

Most, if not all, of the tooling used in production should be placed in a preventive maintenance program that specifies the intervals between maintenance inspections. Ensure that any equipment requiring calibration is included in this program, along with the interval required for calibration. Depending on the complexity of the product and the stage at which production is performed, final inspection can be used to validate the production equipment and tools. However, if a component is hidden or inaccessible at final inspection, you must be able to show that it was produced on equipment that was validated prior to production. Make sure there is a link between the software program included on the drawing and the software program within the CNC machine.

A lockout program that prevents a different software program from being used is the best defense against the misuse of a software program. Be prepared to show your auditor how the lockout program works and to demonstrate inspection processes from the first article produced under the CNC program. Develop a form that ties a software program to the representative first article.

If you have equipment or tooling in storage, implement a procedure to periodically assess its condition.

SECTION 7.5.1.3: CONTROL OF MAINTENANCE EQUIPMENT, TOOLS, AND PROGRAMS (AS9110)

Discussion

The most important part of this subsection, and perhaps the most potentially damaging, is found in the first paragraph, which requires that all maintenance equipment, tools, and programs be validated prior to use. There is a particular sensitivity within the FAA concerning locally manufactured tooling. FAR Part 145.09 at (c) requires the following: "The equipment, tools, and material must be those recommended by the manufacturer of the article or must be at least equivalent to those recommended by the manufacturer and acceptable to the FAA." This sensitivity is a result of accidents that have occurred as a result of improper tooling.

Unusual aspects of this particular subsection include the requirement that periodic preservation and condition checks be established for maintenance equipment or tooling in storage. Registrars will seek to ensure that any equipment put in storage is given a periodic preservation check pursuant to established procedures. CNC equipment and programs must be shown to be validated prior to use.

Findings

Findings are issued under this subsection when maintenance equipment and tools are not validated prior to use, in the case of locally manufactured tools and equipment. The CMM will state a part number and description for special tools and fixtures used to repair and overhaul parts and components. Many times an organization will seek to manufacture these special tools. In the event that the organization chooses to manufacture these tools and fixtures locally, there must be a full showing that the locally manufactured tool is equivalent in all respects to the OEM-mandated special tooling. Major findings are written when such an equivalency cannot be shown.

Best Practices

Most of the locally manufactured tooling used in a repair station is in the nature of fixtures and generic tooling and therefore not unique to a specific manufacturer's product or process, or the CMM or OHM. These tools tend to be cradles, tweezers, standard wrenches, or magnifying glasses. The FAA has determined that equipment that is not "special" in nature need only be designed and calibrated to make measurements within the specific manufacturer's tolerances to be considered equivalent for those tests or measurements. Special tooling such as specific holders, wrenches, or extractors or test equipment designed to measure extremely narrow and specific measurements to a manufacturer's product or process needs to be calibrated and given an equivalency evaluation. If you have equipment or tooling in storage, make sure that you have a procedure to periodically assess the material for condition. The standard is silent on the frequency, and yearly would not be inappropriate.

SECTION 7.5.1.4: POST-DELIVERY SUPPORT

Discussion

Section 7.5.1.4 replaces the requirements that used to be in section 7.5.1.5. This aligns with the changes in ISO 9001:2008 that required post-delivery support activities. The old section was labeled "Servicing," and the new section is labeled "Post-Delivery Support." Some of the changes indicate that it will be almost impossible to exclude this section without an extremely well-written justification.

Findings

Findings related to section 7.5.1.4 are often the result of a failure to properly service or repair parts and components. Generally, the service department handles these items, but it's not uncommon for rework to be done in manufacturing. Sometimes there are discrepancies in this process, such as missing documents (e.g., certificate of conformity, release documents, or test results).

Best Practices

This is an area that most parts manufacturers can exclude from their registrations. Make sure that your organization has a specific procedure on materials returning from *customers* and the materials' approval and return into the system. For example, a component that is returned for updating a printed circuit board must have a certificate of conformity to the exact process that was requested on the purchase order. If the request was for a Mil-Spec process and the certificate of conformity contains a representation of performance to an ASTM specification, this is a discrepancy that should be caught by the receiving inspector and handled accordingly.

AS9100C incorporates eight requirements for post-delivery from ISO 9001. Recent interpretations by the ANSI-ASQ National Accreditation Board (ANAB) and the International Aerospace Quality Group (IAQG) indicate that it will be very difficult to exclude section 7.5.1.4. This is particularly aimed at section 7.5.1.4 (b). The general consensus is that the warranty process is specifically covered under this section. However, it's still possible to exclude the remaining subparagraphs if it can be substantiated or justified.

SECTION 7.5.2: VALIDATION OF PROCESSES FOR PRODUCTION AND SERVICE PROVISION*

Discussion

A common misconception concerning the special process requirements of section 7.5.2 is that they apply to all special processes, wherever they're performed. This is not what this subsection was intended to do. If an organization performs a "special process," section 7.5.2 does apply, but it doesn't apply to special processes that are outsourced, such as plating, nondestructive testing (NDT), painting (in some cases), welding, and soldering (in some cases).

Often, the only way to verify that a particular process has been performed according to specifications is by destroying the product produced. The only suitable substitute for destructive testing is through compliance with accepted standards for NDT. These standards and their related processes are referred to as "special processes."

AS91xx requires that when a special process is used, it must be validated through its full range of requirements. Section 7.5.2 requires the organization to establish criteria for review and approval of the process, which would include qualification and approval of a special process prior to use. Many of these special processes are common within a repair station setting.

The standard requires that if an organization employs NDT as part of its processes, it must define the criteria for review of the process. A second requirement is that the organization arrange for the equipment required to perform the approval, and ensure the qualification of personnel necessary to perform the special processes. A good example of this would be the wet method of dye-penetrant inspection. Usually, written procedures are employed.

*Section 7.5.2 does not apply to AS9120.

Many of these processes can be taken directly from an organization's drawing or process specification. Although section 7.5.2 refers to section 4.2.4 for its record-keeping requirements, most NDT processes also require record keeping.

Under AS9110, many of these processes are called out directly from the CMM, OHM, or even an airworthiness directive or service bulletin. The requirement for record keeping does not mandate a specific AS9110 record keeping process for NDT, despite section 7.5.2 paragraph d. The reference to AS9110 section 4.2.4, Control of Records is overridden by the process in use. Most, if not all, NDT processes require record keeping and certainly anything called out in the CMM, OHM, airworthiness directive, or service bulletins, which have their own specific record-keeping requirements. Make sure you don't build in conflicting requirements for record keeping.

Findings

Many findings are written against subsection 7.5.2 because it is integral to the processes used by the repair station and because of the inability to measure the product directly, for example, an X-ray. You can't look at the product and determine whether it was x-rayed or has internal damage. You must rely on a report from a qualified special processor (whether internal or external), the written certificate of inspection and results, or X-ray film. Without these records, you cannot justify compliance with the special process. Auditors take special interest in auditee organizations' methods, procedures, records, equipment approval, operator certification, and revalidation. Major findings are written when operators perform operations for which they are not certified. Failure to keep records constitutes the greatest source of findings against section 7.5.2. Another large source of findings comes from the organization's failure to ensure that its operators receive their required yearly colorblindness tests or visual acuity tests.

Sometimes, organizations don't maintain records of their required monthly, weekly, and daily tests before they start their NDT. This is an area that the organization should watch out for, as it's fertile ground for nonconformities.

Best Practices

Organizations must have written procedures that cover the implementation of their NDT programs. These procedures cross many requirements of the standard, and process auditing will follow those that impinge on other clauses. This is illustrated in Appendix D, which is a related example procedure for a company that provides fluorescent penetrant inspection.

Remember that training issues have specific applications to personnel who conduct in-house special processes.

SECTION 7.5.3: IDENTIFICATION AND TRACEABILITY

Discussion

One of the cornerstones of an AQMS is the identification and traceability requirement. Organizations must identify their products by suitable means "throughout product realization." It's also a cornerstone of FAA and OEM requirements for parts and materials installed and used in aircraft. Organizations must comply with regulatory authorities. The requirement for maintaining the identification of the configuration of the product to identify any differences between it and the configuration specification is an integral part of the document control system and configuration management system. This is very important when handling several numbered components of the same part number. Take the Lycoming O-360 engine as an example. There are 167 different models within the O-360 family of engines, with 12 different prefixes. This includes the following configurations of the same basic engine:

- O-360 carbureted horizontally opposed engine series
- HO-360 horizontally mounted engine series for helicopter installation
- LO-360 (identical to the O-360 model but with the crankshaft rotating in the opposite direction for use on twin-engine aircraft)
- TO-360 turbocharged engine series
- LTO-360 turbocharged left-hand rotation engine series
- IO-360 fuel-injected horizontally opposed engine series
- LIO-360 (identical to the IO-360 model but with the crankshaft rotating in the opposite direction)
- AIO-360 inverted mount, fuel-injected engine series
- AEIO-360 aerobatic, fuel-injected engine series
- HIO-360 horizontally mounted, fuel-injected engine series for helicopter installation
- LHIO-360 left-hand rotation, fuel-injected, horizontally mounted helicopter engine
- TIO-360 turbocharged and fuel-injected engine series

AS9120 has a specific requirement and note:

> The organization shall maintain product identification and traceability by suitable means (e.g., labels, bar codes) from receipt; during splitting, storage, packaging, and preservation operations; and until delivery (including subcontracted handling or packing operations).
>
> NOTE: Traceability requirements can include
>
> - identification to be maintained throughout the product life,
> - the ability to trace all products manufactured from the same batch of raw material, or from the same manufacturing batch, to the destination (e.g., delivery, scrap),
> - for air assembly, the ability to trace its components to the assembly and then to the next higher assembly, and
> - the identification and condition (e.g., new, repaired, altered or rebuilt) of product in inventory.

Also, AS9120 includes the specific requirement that the organization develop a documented procedure that provides for maintaining the manufacturer's standard hardware that are later sold piecemeal; the ability to identify and trace the products manufactured from the same batch of raw materials or from the same manufacturing batch; and the ability to trace the product to the ultimate destination (delivery, scrap). This means that the product is covered from the cradle to the grave for as long as your company has the product, as well as back in time to the manufacture and raw material sources. The requirement for maintaining the identification of the configuration of the product in order to identify any differences between the actual configuration and the agreed configuration is an integral part of the document control system.

Findings

Split lots and batch identification cause a significant number of findings due to a lack of traceability. Collateral findings occur in this area as well. Traceability problems can cause organizations to miss FAA reporting requirements; thus, one problem can cause a domino effect and ruin an audit. Configuration issues also lead to findings when the agreed configuration (series or revision number) doesn't match the customer's requirements. Repeated findings in this area can jeopardize an organization's AS91xx registration.

Best Practices

The best policy is to generate the required documented traceability program from the onset of business. This applies to all parts of the organization, from purchasing to receiving to warehousing to sales.

The most significant change to this section is the elimination of the requirement for "documented" control for stamps, signatures, passwords, and so on. AS91xx requires "appropriate" controls as an acknowledgment that there are many ways in which media can be controlled and that they should be appropriate to the use, custom, and requirements of the organization.

Despite AS91xx's elimination of this requirement, it's still a good practice to have a process that takes into account the use of stamps, lost stamps, stamp audit programs, and the return of stamps to reduce variation in daily operations. This is especially true if the operation involves flight operations or has FAA oversight.

In AS9110 operations, again, the best policy is to generate the required documented traceability program from the onset of business. This runs across all parts of the organization, from purchasing to receiving to warehousing to sales to shipping to after-delivery activities. Self-disclosures to the FAA should be handled promptly and records of any submissions, replies, and decisions should be kept.

In AS9120, remember that traceability includes condition (new, repaired, rebuilt, and removed).

SECTION 7.5.4: CUSTOMER PROPERTY

Discussion

Organizations are expected to handle customer-owned property carefully and with a full accounting. They are required to report damage or loss of customer property to the customer and maintain records of losses and damage. This section identifies customer-furnished data used for design and maintenance as customer property. This means that the information must be safeguarded and protected just as customers' physical property would be.

In AS9110 the repair station handles customer property almost exclusively. It is expected that the repair station will handle the property with care and in accordance with the manufacturer's specifications. There is also a requirement for reporting damage or loss of customer property to the customer and maintaining records of the loss or damage. This section also identifies customer-furnished data used for design and maintenance as customer property. This means that the customer information must be safeguarded and protected just as any customer property would be handled.

Findings

Mishandled material will result in a finding—for example, shipping containers with specific orientations ("This Side Up") that are stored improperly. This requirement also applies to environmental requirements that are not being met, unused holding fixtures, and metal-to-metal contact. Findings can be written under section 7.5.1 or section 7.5.4, depending on the circumstances. For example, in the case of a steel part left to rust without protection or hydraulic line without caps, the auditor could write a finding under production or customer-owned property requirements. Most findings in this area concern tooling or specialized containers that are damaged or lost.

Best Practices

Ensure that receiving department and inventory management personnel carefully handle materials in accordance with instructions printed on shipping containers, such as "Do Not Stack" and "This Side Up." Avionics shipping containers have special storage and handling requirements, so be sure that a listing of the containers or other product is recorded. Microsoft Excel spreadsheets are commonly used for this purpose. Create a procedure for tracking product loss and damage that includes reporting these statistics to customers. Remember that these reports are uploaded into the OASIS database, and thus customers may cite repeated findings of mishandling customer property as grounds for not doing business.

SECTION 7.5.5: PRESERVATION OF PRODUCT

Discussion

All raw materials, parts, components, and supplies; customer-provided materials or product; and product sent for outsourced work may be damaged, lost, misused, misplaced, or stolen or may become unsuitable (perishable) or obsolete (past shelf life). This could occur during receipt, handling, storage, use, or transportation to the customer; therefore, section 7.5.5 requires the organization to use controls to preserve its products.

The generic portion of this clause calls for general controls, which include inventory cycle counts, stock rotation methods, just-in-time production, shelf-life tracking, material resource planning systems for tracking requirements and use, special controls for restricted access, handling and storage of hazardous materials, climate and environment identification, status and traceability indicators, maintenance procedures, bar codes, training, and use of special equipment for handling.

Although this clause doesn't call for a specific documented procedure, these controls may be included in your processes for sections 7.5.3 and 7.5.4 because many of the controls for those sections also apply to preservation of product.

The specific requirements for cleaning; prevention, detection, and removal of foreign objects; special handling for sensitive products; marking and labeling (including safety warnings); shelf-life control and stock rotation; special handling for hazardous materials; and environmental controls (e.g., temperature, humidity) have been discussed in previous sections of the standard. The few items that haven't been discussed should be incorporated into those earlier processes. Stock rotation (usually associated with first-in-first-out), safety warnings, and hazardous materials need to be addressed. The exception to this is the requirement of the applicable authority or the customer for documents to be protected against loss and deterioration.

Findings

The findings in this area usually stem from a failure to follow existing procedures. Storage (environmental) or shelf-life violations are the most common types of findings. The auditor may write the finding against section 7.5.3 or section 7.5.5, depending on the circumstances. Documentation in the shipped goods can also raise a finding if it is not properly protected from the elements. Failure to ensure that serviceable parts are physically segregated from unserviceable parts will result in a finding in AS9120 audits.

Best Practices

An organization's internal audit should focus on three sections: 7.5.3, 7.5.4, and 7.5.5. Check several times to ensure that there is no metal-to-metal contact or potential FOD issues. Specific customer packaging requirements must be spelled out in a written work instruction, and the packaging must be stored according to these instructions. Complex packaging instructions typically seen on military packaging may be reduced to coded letter groups. If the organization doesn't know these codes, it should seek the services of someone who can translate them into a written work instruction.

SECTION 7.6: CONTROL OF MONITORING AND MEASUREMENT EQUIPMENT

Discussion

The required monitoring and measurement equipment used at the organization must be calibrated, handled, and stored very carefully. The required recall system must be able to tell the company when calibration is due. The organization must maintain a register of calibrated devices and define the process employed for their calibration, including equipment type, unique identification, location, frequency of checks, check method, and acceptance criteria. If the organization outsources the calibration, it must ensure that the calibrating organization is qualified for the calibration and that it uses accepted calibration procedures. If calibrations are performed internally, work instructions are required and the organization must ensure that environmental conditions are suitable for the calibrations, inspections, measurements, and tests.

If a calibration is found to be out of tolerance, the organization must analyze the effect of the calibration error on already-stored or delivered products. For example, if a freezer wasn't maintained at the correct temperature, the organization must notify its customers of the nonconformance. If computers are used to maintain calibration, the organization must validate the computer program and reconfirm it regularly (e.g., temperature or humidity parameters). The standard also requires that employees' personal equipment and customer-supplied equipment used to provide evidence of product conformity be included in the tool calibration program.

Findings

This section is the source of many findings. Almost all of them involve measuring devices that are past their scheduled calibration dates. The lack of adequate recall records is also a common source of findings.

Occasionally, the lack of an analysis for a monitoring and measuring device that's significantly out of tolerance will provide a finding. Tools are often not stored properly, which is another common finding. Employee-owned micrometers, calipers, and other measuring equipment used for final inspection purposes are often not included on the required recall system or aren't correctly calibrated. This can also be the source of many findings.

Best Practices

Identify all monitoring and measuring devices used in the organization—especially those that record continually, such as temperature and humidity chart recorders. There are many excellent metrology software recall programs, such as GAGEtrak, that can be installed quickly and efficiently.

Organizations must have efficient recall programs to avoid missing calibration dates. Prepare a contingency plan to notify customers in the event that an out-of-tolerance condition has led to a questionable product. This should be a smoothly operating process to ensure compliance with product specifications. If the organization outsources its calibration processes, it should have a current copy of the provider's third-party certification (usually to ISO 17025) to ensure compliance with the traceability requirement of the National Institute of Standards and Technology (NIST). The best practice is not to allow employees the option of using their personal measuring equipment for maintenance functions unless it is calibrated along with company-owned tools and equipment.

The aerospace-specific clauses have been amended to allow tools—such as go/no-go gages, for which only verification is available—to be comfortably included in the section on control of monitoring and measuring equipment.

Finally, software confirmation has been given some guidelines with the last note in revision C. If anything, this section's requirements are better defined and more clearly stated.

Be mindful of your responsibilities under section 7.4.1, which requires organizations to evaluate and select suppliers on the basis of their ability to supply product in accordance with the organization's requirements. In short, if you don't have the skill to evaluate tool calibration, you must rely on a third-party certification scheme (such as ISO 9001 or ISO 17025). Section 4.1 of AS9100C clearly mandates the type and extent of control required for outsourced processes, and this must be included in the quality management system (QMS). Ultimately, the responsibility for quality is on the organization that chooses to outsource.

In addition, the new note in section 7.4.1 of AS91xx points out that supplier quality data from objective and reliable external sources (which includes accredited certification bodies) can be used during supplier selection and evaluation. Remember that the use of these data is only one component of an organization's supplier control process and that the organization remains responsible for verifying that purchased products meet specified purchase requirements.

NOTES

Chapter 10
Measurement

SECTION 8.1: MEASUREMENT, ANALYSIS, AND IMPROVEMENT—GENERAL

Discussion

This section requires organizations to determine the extent of their use of statistical techniques, including quality management. The use of these processes must demonstrate product conformity and the health of the aerospace quality management system (AQMS), and show improvements to it.

Findings

Few findings arise in this section, but a lack of data showing the statistical graphs and charts will lead to further examination of the required determination process. This may lead to findings if not substantiated. It is understood that a newly certificated organization may not have the data to support statistical charts at the onset, but with no demonstration of plans to gather these data, a finding will be raised.

Best Practices

Try to align your quality objectives (addressed in section 5.4.1) with the requirements in section 8.1. Planning will become much easier when you aim for an on-time delivery target, track deliveries over time, and review them at management review. The charts need not be anything more than a Microsoft PowerPoint presentation and Microsoft Excel charts that show the target versus the track record for a determined period. Other possibilities are customer returns, return material authorizations (RMAs) versus deliveries, and customer ratings (scorecards). Even employee satisfaction makes a good candidate for statistical presentation.

SECTION 8.2: MONITORING AND MEASUREMENT, SECTION 8.2.1: CUSTOMER SATISFACTION

Discussion

Section 8.2.1 received significant additions because it defines mandatory measures the organization must monitor.

Organizations' customer satisfaction and management review processes must be expanded to include all of revision C's new requirements, not just on-time delivery and customer complaints. Most, if not all, of these metrics can be obtained through internal sources, but, as the note points out, customer surveys are a great source. These items must be discussed at the management review meetings, and related charts and data should be presented at the meeting as objective evidence of the organization's progress with customer satisfaction.

This requires the organization to consider the methods for determining customer satisfaction and demands that it monitor customers' perceptions as a measurement of the effectiveness of its quality management system (QMS).

A newly implemented Industry-Controlled Other Party (ICOP) focus surrounds customer audit findings and escape notices. Ensure that your processes include a review of customer complaints and satisfaction information, including formalized corrective actions and escape notices.

This requirement is not limited solely to external customers; it also applies to internal customers.

Findings

Customer satisfaction systems that consistently receive low feedback scores or that aren't maintained will result in findings. Not having a system at all will certainly bring a major finding. However, findings in this area are rare, as most organizations understand the value of satisfied customers.

Auditors will look for customer corrective actions, complaints, and escape notices. A failure to respond to, investigate, or correct deficiencies will result in findings. These may be major if they are systemic or prolific. The International Aerospace Quality Group (IAQG) recently started requiring organizations to show at every audit that they don't have any pending feedback issues on the OASIS database. Contact your registrar for its requirement on this.

Best Practices

Smart organizations continue to link their quality objectives with the required information under section 8.2.1. However, efficiently gathering this information can be difficult. A company may send out 200 individual surveys to its customers and receive less than 2% back. Even then, the responses it receives may have been completed by someone other than for whom the survey was intended. To combat this problem, consider conducting short telephone surveys. Figure 10.1 illustrates a form representatives could use during surveys.

Make sure your quality system addresses the handling, investigation, reporting, and review of customer corrective action requests and customer complaints. Section 8.2.1 contains three elements:

1. Tracking of on-time delivery, quality, customer complaints, and corrective action requests
2. Taking action when deficiencies are noted
3. Evaluating the effectiveness of actions taken to correct deficiencies

Fourth St. Aircraft Repairs 567 Fourth Street Blakely, WY 98027	**How can we improve?**

Name of company:______________________________
Date:__________________
Sales rep:___________________________________

Product Quality	
How would you rate our on-time delivery? 1 to 5 _______	How would you rate our customer service? 1 to 5 _______
How would you rate our after-delivery service? 1 to 5 _______	How would you rate our pricing? 1 to 5 _______
How would you rate our documentation? 1 to 5 _______	

Additional comments: ____

Thank you for your participation!

Figure 10.1 Customer satisfaction survey.

Make sure that your system provides for these elements and that you can show the auditor what you are doing when your on-time delivery slips below your or your customer's quality objectives. An action plan is the ideal place to address these elements. Make sure you follow up on them at the management review.

SECTION 8.2.2: INTERNAL AUDIT

Discussion

One of the premier cornerstones of an AS91xx-compliant AQMS is its self-assessment section. The number of days a registrar must dedicate to auditing a client's QMS is set by two documents: IAF MD 5:2009 and AS9104/1. This calculation is made using the number of employees in the organization—including temporary and part-time employees—along with other factors. The ICOP scheme has criticized this method of audit delivery on the basis that auditors don't spend enough time at auditee sites during annual audits. This means that auditors must rely on the effectiveness and reliability of their clients' internal audit systems to supplement their observations during on-site audits. Each visit should include a review of the internal audit schedule, audit reports, auditor qualifications, and findings that were discovered during the organization's internal audits.

There is a very different philosophy between the audits performed by a third-party registrar and internal audits. Third-party auditors always strive to find evidence of conformity. Internal auditors tend to focus on evidence of nonconformity. This improves the system and is an essential part of the continual improvement process. Internal audits are a form of preventive action. It's important that the organization understand that a third-party auditor finding no nonconformances in internal audit reports could be a serious problem.

AS91xx requires that organizations plan an audit program at designated intervals. The intervals and scopes of these audits are not stated but are left to the organization to develop. The only guidance given is that the organization must consider the status and importance of the processes and areas to be audited, as well as the results of previous audits. Does this mean that everything in the system must be audited annually? Possibly, but even if an organization internally audits each of its functions annually, its third-party auditor will likely audit each one again.

This section contains numerous requirements that must be addressed. A documented procedure is required that covers the responsibilities and requirements for planning and conducting audits, and for reporting results and maintaining records. This procedure must describe the audit criteria, scope, frequency, and methods used in audits.

Auditors must be selected to ensure impartiality and objectivity and must not audit their own work. Auditor qualifications must be established. Section 6.2.1 states: "Personnel performing work affecting product quality shall be competent on the basis of appropriate education, training, skills, and experience."

Section 6.2.2 states: "The organization shall determine the necessary competence for personnel performing work affecting product quality." Training must be provided that shows competence with AS9120 section 8.2.2 (a): "The organization shall conduct internal audits at planned intervals to determine whether the quality management system conforms to the planned arrangements . . . [and] to the requirements of this International Standard and to the quality management system requirements established by the organization."

When findings are issued, there is a special requirement that differs from the corrective action process requirements. Section 8.2.2 provides that "the management responsible for the area being audited shall ensure that actions are taken without undue delay to eliminate detected nonconformities and their causes. Follow-up activities shall include the verification of the actions taken and the reporting of verification results (see section 8.5.2)."

Finally, the audit program must meet contractual or regulatory requirements. Under European certification rules, the European Aviation Safety Agency (EASA) requires an annual internal audit. The use of checklists for audits is no longer mandated by the standard, but it's highly recommended that checklists or other processes be refined to assist the organization as part of its internal audit program.

Findings

Section 8.2.2 is a common source of findings. For example, auditor qualifications may not be defined or documented. The use of untrained or unqualified internal auditors constitutes a finding. Failure to meet the audit program frequency or keep to the audit schedule is a common finding; failure to perform a substantial number of internal audits constitutes a major finding. Prior findings that are corrected with "undue delay" are a common finding, as are audit reports that are incomplete or include data that aren't supported by objective evidence.

Internal auditing is one process you want full and complete compliance with.

Best Practices

The introduction to AS91xx encourages the process approach:

> This International Standard promotes the adoption of a process approach when developing, implementing, and improving the effectiveness of a quality management system, to enhance customer satisfaction by meeting customer requirements.
>
> For an organization to function effectively, it has to identify and manage numerous linked activities. An activity using resources, and managed in order to enable the transformation of inputs into outputs, can be considered as a process. Often the output from one process directly forms the input to the next.
>
> The application of a system of processes within an organization, together with the identification and interactions of these processes, and their management, can be referred to as the "process approach."

An advantage of the process approach is the ongoing control that it provides over the linkage between the individual processes within the system of processes, as well as their combination and interaction.

When used within a QMS, such an approach emphasizes the importance of the following:

- Understanding and meeting requirements
- The need to consider processes in terms of added value
- Obtaining results of process performance and effectiveness
- Continual improvement of processes based on objective measurement

Registrars emphasize the use of the process approach to auditing. An organization's audit plan should ensure that all applicable sections of the standard are covered. Top management should discuss the definitions of various internal processes to determine which represent the widest view of the inputs and outputs. Figure 10.2 illustrates a chart to help determine this.

Department and function audits should be scheduled based on the results of previous audits and the importance of the department or function. Figure 10.3 illustrates a sample audit schedule. Notice that in the example, document control is tagged as especially important, so it's scheduled to be audited twice in the sample cycle.

Whatever system you use, simplify it as much as possible. Avoid complicated processes for approving plans, audit reports, and the like.

Finally, auditor training should be carefully recorded and performed. Find a good online internal auditor course. Registrars often offer them for a nominal fee. Retain a copy of the attendance certificate for training records, as well as for the auditor's wall.

	1	2	3	4	5	6	7	8	9	10
Process name	**Core management**	**Purchasing**	**Calibration**	**Contracts**	**Nonconforming product**	**Internal audit**	**Document/record control**	**Resource management**	**Production line 1**	**Production line 2**
4.1 General requirements	✗									
4.2.1 Documentation requirements—general	✗									
4.2.2 Quality manual	✗									
4.2.3 Control of documents	✗							✗		
4.2.4 Control of records	✗							✗		
5.1 Management commitment	✗									
5.2 Customer focus	✗									
5.3 Quality policy	✗									
5.4.1 Quality objectives	✗									
5.4.2 Quality planning	✗									
5.4.3 Safety objectives (AS9110 only)	✗									
5.5.1 Responsibility and authority (including 5.5.1.1–2 for AS9110 only)	✗									
5.5.2 Management representative	✗									
5.5.3 Internal communication	✗									
5.6 Management review (includes 5.6.1, 5.6.2, 5.6.3)	✗									
5.7 Safety policy (AS9110 only)	✗									
6.1 Provision of resources								✗		
6.2.1 Human resources—general								✗	✗	✗
6.2.2 Competence, training, and awareness								✗	✗	✗
6.3 Infrastructure									✗	✗
6.4 Work environment									✗	✗
7.1 Planning of product realization	✗								✗	✗
7.1.1 Project management (N/A for AS9120)	✗									
7.1.2 Risk management (N/A for AS9120)	✗									
7.1.3 Configuration management (7.1.1 for AS9120)	✗				✗				✗	✗

Figure 10.2 Internal QMS processes.

Process name	1 Core management	2 Purchasing	3 Calibration	4 Contracts	5 Nonconforming product	6 Internal audit	7 Document/record control	8 Resource management	9 Production line 1	10 Production line 2
7.1.4 Control of work transfers (7.1.2 for AS9120)	✗								✗	✗
7.2.1 Determination of requirements related to the product	✗		✗							
7.2.2 Review of requirements related to the product	✗		✗							
7.2.3 Customer communication	✗		✗							
7.3.1 Design-and-development planning (N/A for AS9120)										
7.3.2 Design-and-development inputs (N/A for AS9120)										
7.3.3 Design-and-development outputs (N/A for AS9120)										
7.3.4 Design-and-development review (N/A for AS9120)										
7.3.5 Design-and-development verification (N/A for AS9120)										
7.3.6 Design-and-development validation (including 7.3.6.1–2) (N/A for AS9120)										
7.3.7 Control of design-and-development changes (N/A for AS9120)										
7.4.1 Purchasing process	✗									
7.4.2 Purchasing information	✗					✗				
7.4.3 Verification of purchased product		✗			✗		✗		✗	✗
7.5.1 Control of production and service provision (including 7.5.1.1–4) (N/A for AS9120)							✗	✗	✗	✗
7.5.2 Validation of processes for production and service provision	✗		✗				✗	✗	✗	✗
7.5.3 Identification and traceability	✗				✗		✗	✗	✗	✗
7.5.4 Customer property	✗				✗				✗	✗
7.5.5 Preservation of product					✗				✗	✗
7.6 Control of monitoring and measurement equipment			✗				✗	✗	✗	✗

Figure 10.2 Internal QMS processes. *(Continued)*

(continued)

Process name	1 Core management	2 Purchasing	3 Calibration	4 Contracts	5 Nonconforming product	6 Internal audit	7 Document/record control	8 Resource management	9 Production line 1	10 Production line 2
8.1 General	✗									
8.2.1 Customer satisfaction	✗		✗							
8.2.2 Internal audit			✗							
8.2.3 Monitoring and measurement of processes	✗									
8.2.4 Monitoring and measurement of product	✗									
8.2.5 Evidence of conformity (AS9120 only)	✗		✗							
8.3 Control of nonconforming product										
8.4 Analysis of data	✗					✗	✗			
8.5.1 Continual improvement										
8.5.2 Corrective action						✗	✗			
8.5.3 Preventive action										

Figure 10.2 Internal QMS processes. *(Continued)*

2014 Audit Schedule					
Process	**Month**	**Auditor**	**Completed**	**Report complete?**	**Findings**
Management processes	January	Smith	1/12/2014	Yes	1
Purchasing process	February	Jones			
Quality processes	March	Jones			
Receiving inspection process	April	Jackson			
Warehouse processes	May	Smith			
Documentation control	June	Jackson			
Management processes	July	Jones			
Competence, awareness, training	August	Smith			
Customer-related processes	September	Jones			
Documentation control	October	Smith			

Figure 10.3 Sample audit schedule.

SECTION 8.2.3: MONITORING AND MEASUREMENT OF PROCESSES

Discussion

Section 8.2.3 is a reminder to ensure that process outputs are monitored. Are the proper documents available to provide objective evidence that all maintenance operations have been completed as planned? Does the completed maintenance activity conform to Federal Aviation Administration (FAA) requirements? Is the correct configuration delivered to the customer? These processes are among the many other output processes that need to be monitored.

The standard requires that suitable methods be applied. Monitoring and measurement can take the form of rejections from customers, FAA violations, or customer complaints on delivery conditions. They can also be direct measurements, such as dock or product audits of departing products. Regardless of the method, organizations must take corrective action to correct the process, evaluate whether nonconforming goods have been delivered, and identify and control any nonconforming product.

AS91xx makes it clear that organizations must determine whether a process nonconformity is limited to a specific case or whether it could have affected other processes or products. This will cause the auditor to look at the analysis and results of process nonconformities. This should be included in the organization's corrective action forms and process.

Findings

There are few findings related to this section. Of those that are found, most stem from a failure to follow the organization's procedure for monitoring and measurement of processes. Another area concerns the control of nonconforming product already delivered to the customer.

Best Practices

The implementation of audits after final inspection at frequent intervals is the best source of monitoring to avoid customer complaints. Rotate various personnel to perform these audits. Train supervisors and technicians to perform a simple audit of completed projects on a calendar basis (e.g., weekly). Report these findings as part of the standard's analysis of data section (8.4.1) and discuss them at management reviews. Ensure that the corrective action program funnels through this process.

SECTION 8.2.4: MONITORING AND MEASUREMENT OF PRODUCT

Discussion

The standard requires the use of a "statistically valid and appropriate for use" sampling plan. Many firms use MIL-STD-105A for this purpose, and this is acceptable. Standard military sampling procedures for inspection by attributes were developed during World War II. US Army ordnance tables and procedures were generated in the early 1940s, and those grew into the army service forces tables. The result of these programs was MIL-STD-105A. It has been modified from time to time and issued as 105B, 105C, and 105D and was adopted in 1971 by the American National Standards Institute (ANSI) as ANSI standard Z1.4 (which is also an acceptable sampling standard). In 1974, this standard was adopted (with minor changes) by the International Organization for Standardization (ISO) as ISO 2859. The latest MIL-STD-105 revision is MIL-STD-105E and was issued in 1989.

There is a new requirement that processes must be in place to control and monitor critical items and key characteristics. When critical items have been identified, the organization shall ensure that they are monitored and controlled.

AS9110 has a special requirement for the organization to identify defects discovered during maintenance that are outside the scope of the maintenance contract and to process them in accordance with customer and authority requirements.

The standard also requires that procedures be implemented for the preparation and completion of authority documentation, including release documentation and approvals for return to service. The balance of this section requires records of conformity and the identification of the person authorizing release of the product to the customer.

If the product is discovered to have damage or a nonconformity that is outside the scope of the maintenance activity, this must be reported to the customer and, if necessary, to regulatory authorities.

Findings

Findings in this area usually involve a failure to identify who releases product to the customer. Stamps are perfectly satisfactory, as long as each individual has his or her own stamp. Findings involving incorrect sampling sizes are also common. This may be attributed to training or to the lack of a valid work instruction for the sampling process. Other potential findings involve incorrect preparation of product release documentation or a lack of evidence of conformity of product maintained and delivered to a customer.

Best Practices

A stamp roster and procedure to identify people authorized to release product are a must in this area. Provide the type of stamp that will be used (i.e., acceptance, rejection, release), and include a procedure in the event that a stamp is lost. Make sure that the Suspected Unapproved Parts (SUP) program is defined and fully functioning. Establish the sampling plan and ensure that individuals are trained in the use of these tables and sample size selection. If an organization uses copies of sample tables as posters in the work area, their source and revision level must be clearly shown. The FAA has issued an order detailing the preparation of FAA form 8130.3, and this should be used as part of the required procedure. The malfunction and defect (M&D) report and system difficulty report (SDR) process is a requirement for the FAA-required repair station manual/quality policy document and should be highlighted in the AQMS, even if it is just a reference.

SECTION 8.2.4.1: INSPECTION DOCUMENTATION (AS9100)

Discussion

This section describes the required documentation that should be generated and maintained as part of the maintenance and operation of the repair station. Most of the requirements are contained in the respective component maintenance manual (CMM) or overhaul manual (OHM) in the tables and limits section. This section requires that any testing performed be recorded along with the test data, when required by the specification or test plan. This includes hydraulic testing, engine run-up testing, and electrical component testing, among others.

Findings

As long as the receiving inspection program and the production documented procedures are followed, few findings will be raised. Findings are raised when the documentation for the testing plans is omitted. Most findings will be written against other sections rather than against this section since the other sections tend to occur earlier in the process chain.

Best Practices

Make sure that your documented procedures have records that comply with this section specifically—especially if you use monitoring and measuring tools such as comparators, micrometers, calipers, or the like in the inspection process. If you do perform testing, make sure the test plan has an associated record attached to the plan and that it is used and maintained. Ensure that your documentation packages are complete in all respects. This includes a detailed handling of life-limited items. In addition to the regulatory requirements and customer requirements, the organization may pose an additional requirement, such as the Coordinating Agency for Supplier Evaluation (CASE). An example of this is the requirement for the repair station to maintain a scrap record on life-limited items. This is solely a CASE requirement, but if you are a signatory to CASE this requirement must be performed. The registrar will verify this during the audit.

SECTION 8.2.4.1: NEW MAINTENANCE PROCESSES (AS9110)

Discussion

This section requires that any new maintenance process be documented and approved by the customer and/or regulatory authorities. It is straightforward and applies to processes that are developed by the organization. Of course, as discussed earlier, this would be a requirement if your facility is a designated alteration station or if you have a designated engineering representative (DER) or designated airworthiness representative (DAR) on staff. This section has also been read to mean that the organization must validate any new maintenance process that comes from the original equipment manufacturer (OEM) or supplemental type certificate (STC) holder.

Findings

This section has very few findings. Failure to obtain approval for alternate means of compliance (AMOC) has led to findings, but they are very rare.

Best Practices

Ensure that any maintenance process that you use has been approved with "approved data" or "acceptable data." It's always wise to get customer approval to use any process outside the basic CMM or OHM or maintenance manual.

SECTION 8.3: CONTROL OF NONCONFORMING PRODUCT

Discussion

This section takes the corresponding ISO 9001:2008 requirements and adds to them. In addition to the duty to segregate and mark nonconforming products is the requirement to notify customers of any nonconforming product that they have received and accepted. Records of nonconformity must be maintained, including concessions by the customer.

There is a documented procedure required that contains the same ISO 9001:2008 requirements but adds two additional requirements: the authority and responsibilities of persons authorized to make dispositions must be clearly explained, and the procedure to approve the personnel making dispositions must be included. It also requires that, in addition to any contract or authority reporting requirements, the organization's system shall provide for timely reporting of delivered nonconforming product that may affect reliability or safety. Notification shall include a clear description of the nonconformity, which includes, as necessary, parts affected, customer and/or organization part numbers, quantity, and date(s) delivered. This process is inclusive of the SUP program and the reporting process for unapproved parts.

Another change to AS91xx is the addition of containment requirements in paragraph (e). This will require changes in the documented procedure to ensure that containment actions are appropriate for the effect of the nonconformity.

The organization's documented procedure for nonconforming product must include both provisions for customer approval in its matrix and planning for containment actions. Examples of containment actions include stopping a machine from operating when it's discovered that the machine is out of tolerance or when material is defined as the cause of the nonconformity and supplier action is necessary.

The process to handle nonconforming product must be described by one or all of the following ways: by taking action to eliminate the detected nonconformity; by authorizing its use, release, or acceptance under concession by a relevant authority and, where applicable, by the customer; and/or by taking action to preclude its original intended use or application.

In the case of a repair station, it may be that the repair station doesn't own the part that's to be scrapped. It may be customer property, which would conflict with section 7.5.4. This doesn't mean that the part shouldn't be segregated and clearly marked as scrap material. Life-limited parts that are scrapped will require special handling and, in some cases, record keeping. Remember that AS91xx is complementary to customer requirements.

AS9120 makes it clear that the distributor has no authority to rework or repair product and that dispositions are limited to scrap, rejection for return to the supplier, rejection for revalidation by the manufacturer, and submittal to customer and/or design authority for "Use As Is" disposition.

Findings

Findings are common in this area, and they generally relate to nonconforming product that's not clearly marked or segregated. The organization's procedure often doesn't comply with the requirement for the documented procedure to contain the process for approving personnel to make dispositions of nonconforming products. Some findings occur when internal departments deviate from procedures and instead create their own procedures for holding goods without proper documentation. Goods are either "conforming" or "nonconforming" unless a procedure allows such a hold. Failure to list types of dispositions and who can make them is another common finding in this area. Mixing nonconforming with conforming parts or scrap parts with conforming parts is cause for many findings. In AS9120 audits, repairing, assembling, or dispositions outside of the limitation for AS9120 are cause for findings.

Best Practices

Simplify this process as much as possible. Clearly mark areas used for nonconforming products or identify and segregate them with a traveling component. Do not allow the parts to be mixed with serviceable items. To prevent their use by unscrupulous operators, make sure that the organization has a process and a log for handling life-limited parts that are scrapped.

Create a holding procedure that allows a product to be tagged until its paperwork arrives. It may or may not be nonconforming, depending on the content of the paperwork when it arrives. Make sure that there is a training program for the personnel who make dispositions and that it includes the documented procedure for nonconforming product.

For AS9110 organizations, make sure that you have a process for handling customer-owned life-limited parts that are scrapped. This will prevent later use by the customer or by someone to whom the customer disposes the parts, and protects you from liability. Make sure that the organization has a notification process in the event that the customer receives nonconforming product. Make sure you have training in the FAA SUP program and that your personnel are aware of how to make a report to their supervisor or to the FAA, as applicable.

If a large number of products are declared nonconforming and your segregated area cannot hold them all, use yellow caution tape to mark the products so they can't be introduced into the stream of commerce, mixed with conforming products, or shipped to your customer.

NOTES

Chapter 11
Analysis and Improvement

SECTION 8.4: ANALYSIS OF DATA

Discussion

Appropriate data to demonstrate the suitability and effectiveness of the quality management system (QMS) seem to add a new requirement. All along we have been discussing supplier data, training data, calibration data, process validation data, customer feedback and customer surveys, process and product monitoring and measuring, and nonconforming materials. All of these data points must be analyzed.

Findings

Findings are routinely issued if there is no evidence that the organization analyzed the data it collected regarding customer satisfaction, conformity to product requirements, characteristics and trends of products and processes, and suppliers. It's not for the registrar or auditor to decide what to analyze; it's the organization's decision. As long as these areas are covered and objective evidence is available to support them, the organization won't be found to be noncompliant.

Best Practices

There should be a common thread for the required subsections on suppliers, customer satisfaction, monitoring and measuring of products and processes to analyze, and management review. The logical place for the charts, graphs, and statistical methods to be analyzed is in management review. Make these a part of the organization's quality objectives to complete the thread and maximize your effort toward the continual improvement requirement of section 8.5.1.

SECTION 8.5: IMPROVEMENT, SECTION 8.5.1: CONTINUAL IMPROVEMENT

Discussion

This section requires continual improvement of the entire aerospace quality management system (AQMS). Notice that all the large management processes are mentioned by name:

- Quality policy
- Quality objectives
- Audit results
- Analysis of data
- Corrective actions
- Preventive actions
- Management review

The use of the methods discussed throughout this book will lead to compliance with this section. Make sure that each of the tools discussed in each of the mandated sections is implemented.

Findings

Few findings are written directly against this section. If there are nonconformities, they usually relate to the enumerated section rather than continual improvement. However, an organization's failure to use these tools will result in findings—some of which may be major.

Best Practices

The new note in section 8.5.1 includes examples where opportunities for improvement can be identified. This is a minimal list and should not be construed as a limit of improvement opportunities.

Implement the practices discussed at the appropriate section of each of these enumerated areas. Make sure that the management review minutes become a repository of proof for the required data collection, analysis, and reporting. The International Organization for Standardization's ISO 9001 Auditing Practices Group published the following guidance in December 2009:

> The auditor has to verify how the overall corporate objectives have been translated into internal requirements throughout the appropriate processes, and how these requirements are communicated and monitored. So, the auditor should look for evidence that the organization is analyzing data from process monitoring, and is then taking the results forward for evaluating process efficiency and/or improving process output. One point that should be specifically examined is the consistency of the way in which the improvement of any one process contributes to meeting the

overall objectives, in order to ensure that this will not cause conflict in the achievement of other objectives.

ISO 9001 lists a number of areas that an auditor can assess to obtain evidence of both planning and actual implementation of improvement. Examples of areas where the quality management system can be improved include, but are not limited to:

- Internal communications
- Follow-up activities
- Documented procedures
- The effectiveness of management review meetings
- Customer feedback systems
- Training programs (e.g., for management or for internal auditors)

The auditor needs to look for evidence of how the corporate objectives are translated into specific QMS objectives. For example, an organization could set an objective to reduce customer complaints by 30%. The top management analysis shows that 50% of the complaints concern overdue deliveries. The auditor should then look for evidence that the organization is monitoring and analyzing key aspects of its scheduling and planning activities throughout its processes, and the process interfaces, to reduce delays.

It's important to understand that continual improvement doesn't necessarily mean only continual improvement of product or process; it can and should also apply to the QMS itself.

Auditors should be aware that it is unrealistic to expect an organization to make progress on all potential improvements simultaneously. Continual improvement should be interpreted as a recurring (step-by-step) activity. It means that when opportunities for improvement are identified and are justified, an organization needs to decide how they are to be implemented based on the available resources.

Each improvement will require the commitment of resources. This may require prioritization by top management, especially where investments are needed. The auditor should seek to ensure that the improvement objectives are consistent overall and are coherent with the trilogy of factors mentioned earlier. However, an organization that does not have a policy and objectives relating to continual improvement is clearly not complying with the standard. Similarly, the absence of any evidence of improvement on at least one of these aspects indicates that an organization's quality policy is not in line with ISO 9001.

One warning: The organization is not required to set objectives for improvement of all its processes at any one time. As in the previous example on reducing customer complaints, some processes may not be deemed by top management to contribute significantly to the reduction of delays. It's only normal, therefore, that the organization would not concentrate on those areas.

If top management has set a realistic objective for a process and there is no evidence of improvement, this information must be fed back into the management review so that top management can decide what type of action is appropriate (e.g., readjusting the objective or providing other means to improve the process).

SECTION 8.5.2: CORRECTIVE ACTION

Discussion

The existing ISO 9001 requirements have been supplemented with three additional requirements to the documented procedure required by section 8.5.2: (1) flow-down of the corrective action requirement to a supplier when it's determined that the supplier is responsible for the nonconformity, (2) specific actions where timely and/or effective corrective actions are not achieved, and (3) a process to determine whether additional nonconforming product exists based on the causes of the nonconformities.

The first additional requirement is often misunderstood. Many think that this is a reference to nonconforming material (section 8.3) or verification of purchased product (section 7.4.3). Actually, this section refers to a nonconformity report (NCR) issued to an organization and the organization discovers that the problem is a result of malfeasance on the part of one of its vendors. This requires the organization to develop a process to flow down its root cause analysis to its vendor and seek a response.

The second additional requirement concerns the escalation clause for time-sensitive or ineffective corrective actions, both from an organization's internal and external sources. This is a two-pronged issue: time-sensitive and ineffective. There should be processes for both issues.

The standard requires that the organization take action to eliminate the cause of nonconformities to prevent their recurrence. (Remember that section 8.2.2 requires that corrective actions detected during internal audits be handled without undue delay.) The actions must be appropriate to the nonconformity. Paragraph (c) requires that the process contain the methodology for evaluating the need for action to ensure that nonconformities do not recur.

The third requirement is to cross-check across similar processes and products for additional problems. It's quite straightforward, but it must be included in the documented procedure for corrective actions.

Remember that the difference between corrective and preventive actions is that corrective actions are reactive while preventive actions are proactive.

Findings

The many findings written against this section usually relate to overdue corrective action plans. The auditor will make sure that the organization's procedure for corrective actions is followed. If the procedure says that the corrective action must be done in 28 days, then any corrective action open longer than 28 days is subject to a finding. Some findings are written because corrective actions weren't formally entered into the corrective action system. If there aren't procedures for the flow-down of timely and effective issues, then findings are issued.

Repeated corrective actions on a subject may lead to a finding for not taking action to prevent its recurrence. Many findings are written for failure to follow up on the effectiveness of the actions taken. Corrective actions that are closed out yet written with some action pending are also grounds for nonconformity.

Best Practices

Make sure that the organization's escalation clause has a complete loop. Don't refer it to a supervisor or management and fail to require some action to be taken. Remember that there must be a process for ineffective corrective actions. Make this as simple but as effective as possible. The corrective actions themselves should have a solid root cause analysis. Don't restate the nonconformity.

Also, consider employing the "five whys" system, which is illustrated in the following example:

Problem: The forklift won't start.

- Why 1: The battery is dead.
- Why 2: The generator is inoperative.
- Why 3: The generator brushes are unserviceable.
- Why 4: The generator is well beyond the manufacturer's recommended replacement interval.
- Why 5: The preventive maintenance department hasn't been maintaining the forklift according to the recommended service schedule. (This is the root cause.)

This process encourages the troubleshooter to avoid obvious assumptions and to trace the chain of causes in direct increments from the statement of the problem to a root cause that can be traced to the original issue.

Create a corrective action log to track corrective actions: their assignees, their due dates, the dates they're implemented, and their follow-up dates.

Make sure that all corrective action implementation is in the past tense. Don't close out a finding that is written with a future-tense promise to do something.

The only true addition to the requirements of section 8.5.2 concerns the addition of paragraph (j) and the requirement that the procedure include a process for determining whether additional nonconforming product exists based on the causes of existing nonconformities and the subsequent action taken. This will require additional effort on the part of the organization. The corrective action form should contain a link to the organization's actions for determining whether additional nonconforming product exists and whether action is required.

A new requirement in section 8.5.2 mandates that the organization detail in its documented procedures a process for determining whether additional nonconforming product exists based on the causes of the nonconformities. It also requires the organization to take action when necessary.

For AS9110 organizations, the new standard requires that the corrective action process evaluate the need for action on the basis of human factors to ensure that nonconformities do not recur. Make this a part of your corrective action form.

SECTION 8.5.3: PREVENTIVE ACTION

Discussion

This is probably the most misunderstood clause in the standard. The requirements read similarly to those of section 8.5.2, and many think that the preventive aspects of a corrective action are what this section addresses.

Many organizations join the documented procedures required by sections 8.5.2 and 8.5.3 into one document. However, the language of this section speaks of potential nonconformities, not realized nonconformities.

The same requirement for evaluating the need for action based on human factors to prevent occurrence of nonconformities appears in this section as well as section 8.5.2 for AS9110 organizations.

Findings

The lack of any preventive actions is grounds for a finding. If the auditor believes that this lack is systemic, it will be a major finding. The failure to follow up on the effectiveness of preventive action is also a source of many findings.

Best Practices

Create a preventive action procedure apart from the corrective action procedure. There are no time constraints on preventive actions, so joining the processes may put a time constraint on them. Create a preventive action log or use the corrective action log and clearly define the difference between a preventive action and a corrective action. Use a different numbering system. At the management review, follow up on preventive actions taken and make a record of this review.

The requirement that the documented procedure include a provision for "reviewing" in paragraph (e) has been clarified to show that the act of checking whether actions were taken was not the intent of the revision. Rather, the intent of this section is to determine the effectiveness of a subject to achieve established objectives.

Include references to human factors in your preventive action form.

NOTES

Appendix A

FAA Order 8900.1 Change 0

VOLUME 6: SURVEILLANCE

Chapter 11: Other surveillance

Section 20: Evaluate special equipment or test apparatus

6-2641: Program tracking and reporting subsystem (PTRS) activity codes

A. Maintenance: 3626/3658

B. Avionics: 5626/5658

6-2642: Objective

This section provides guidance for evaluating special equipment and test apparatuses used by a certificate holder in maintaining aircraft and their associated components.

6-2643: General

A certificate holder must make a determination of acceptability for equivalency of special equipment and/or test apparatuses used in maintaining aircraft and their associated components.

A. The term "equivalency," as used throughout this section, means equivalent to that recommended by the original equipment manufacturer (OEM) for the purpose of performing specific tests or making required measurements to determine airworthiness. To determine equivalency, a certificate holder should compare the required test operations or specifications and the technical data of the special equipment or test apparatus (both the data recommended by the manufacturer and the data proposed by the certificate holder). The special equipment or test apparatus may look different, be made of different materials, be a different color, etc. However, as long as the tool is functionally equivalent for the specific test application, the tool may be used.

B. The accuracy of special equipment or test apparatus used to perform a specific task should be at least equal to that recommended by the manufacturer.

6-2644: Background

A. The use of equivalent equipment was first addressed in the Civil Aeronautics Manual 52 (CAM 52), repair station certificates, which preceded Title 14 of the Code of Federal Regulations (14 CFR) part 145. Specifically, CAM 52, section 52.30-1 stated, "The applicant will have the responsibility of choosing suitable tools and equipment (which may be either equipment or tools recommended by a manufacturer in the overhaul or repair of his product or the equivalent of such equipment or tools). . . . The inspecting civil aviation authority (CAA) agent will determine if these tools, equipment, and materials are satisfactory within the intent of this regulation." This implied that the CAA inspector was to make a determination of equivalency. However, this statement simply established that the CAA inspector determined if the equipment was satisfactory. The intent remains the same today. Today, 14 CFR part 43, section (§) 43.13(a) states, "If special equipment or test apparatus is recommended by the manufacturer involved, he must use that equipment or apparatus or its equivalent acceptable to the administrator."

B. A finding of equivalency can only be made based on an evaluation of a technical data file. Additionally, demonstrating functionality of the special equipment or test apparatus may sometimes be required. A technical data file may include, but is not limited to, data, drawings, specifications, instructions, photographs, templates, certificates, and reports. For calibration equipment, the technical data file should also include data sheets attesting to its accuracy when calibration standards are necessary. This file should also describe any special manufacturing processes that are used in the controlling processes, including gauges and recording equipment. If calibration equipment is involved, documented procedures will be used to evaluate the adequacy of the calibration equipment, and the equipment must be traceable to the National Institute of Standards and Technology (NIST) or to a standard provided by the equipment manufacturer. With foreign equipment, a standard of the country of manufacture may be used if found acceptable by the FAA.

C. Most of the test apparatuses used for making airworthiness decisions are generic in nature and designed to make measurements that are not unique to a specific manufacturer's product or process. Equipment that is not "special" in nature only needs to be designed and calibrated to make measurements within the specific manufacturer's tolerances to be considered equivalent for those tests or measurements.

Note: The FAA has allowed the industry to fabricate its own equipment and/or apparatus (test fixtures, jigs, tooling, etc.) without much FAA assessment. Many certificate holders, especially those which employ engineering departments, are very well versed in fabricating special equipment or test apparatus. In some cases, the fabricated special equipment or test apparatus has exceeded the component manufacturer's requirements for accuracy, reliability, etc.

1. Occasionally, a tool that was manufactured by an air carrier did not always produce the same results as the tool from the OEM. Further,

previous industrywide validations revealed that some tools were not equivalent in function to the OEM's specifications.

2. Some tools had been manufactured by a method known as reverse engineering. Reverse engineering alone, without data, drawings, testing, or reports, may not adequately produce a tool or fixture functionally equivalent to an OEM's requirements.
3. With recent technological advances, highly specialized test equipment or test apparatuses are frequently required. Use of such equipment supports the continued airworthiness of aircraft systems and components to the manufacturer's specifications and tolerances.
4. Determining the equivalency of equipment and/or apparatus is the primary responsibility of the certificate holder, not the FAA. The basis of equivalency for equipment or apparatus for products being maintained must meet the OEM's standards and specifications for tolerances and accuracy.

6-2645: Coordination requirements

This task may require coordination with other specialties, regions, or district offices.

6-2646: References, forms, and job aids

A. References: 14 CFR parts 43, 121, 135, and 145

B. Forms: None

C. Job aids: None

6-2647: Procedures

A. Because it is the primary responsibility of the certificate holder, the FAA principal inspector (PI) may only make an acceptance of functional equivalency for special equipment or test apparatus. The following procedures are listed as a reference for making this determination:

1. Ensure that the limitations, parameters, and reliability of the equipment allow it to function equivalently to the OEM's recommended equipment for the specific test application. This may include the review of data from the OEM or another source of data used to manufacture the equipment.
2. Compare the technical data used to that of the manufacturer (if obtainable; manufacturers are reluctant to release technical information about tools and test equipment) and that used by the certificate holder. If the OEM technical data is not available, then the certificate holder must perform an evaluation to make a determination of functional equivalency. If needed, observe test demonstrations of the equipment.
3. Ensure that specific instructions pertaining to the proper use of any special equipment or test apparatus are provided in and adequately referenced in the repair station's inspection procedures manual (IPM)

or the air carrier's continuous airworthiness maintenance program (CAMP).

4. Ensure that the certificate holder has included any special equipment or test apparatus that requires inspection and/or calibration in their test inspection and calibration programs. They should also address the regular inspection and calibration of any special equipment or test apparatus used for making final airworthiness determinations.
5. The FAA recommends that the PI review the repair station's IPM or the air carrier's CAMP to ensure that it has adequate procedures in place, if applicable, for manufacturing and/or determining equivalency of any special equipment or test apparatus in use.

   Note: An aircraft engineering division (AIR-100) memorandum dated December 21, 1999, states, "Designated engineering representatives (DER) may not approve or determine equivalency of tooling and test equipment." Furthermore, neither the FAA nor a DER may approve equipment and/or a test apparatus. The FAA and DER may only make an acceptance of functional equivalency for special equipment or a test apparatus. It is important to emphasize that the burden of demonstrating equivalency is borne by the certificate holder and not the FAA.
6. If unable to make a determination of equivalency from the technical data or through testing, the PI will coordinate with the aircraft evaluation group with respect to the appropriate aircraft/rotorcraft/engine certification office to assist in determining equivalency.

B. Standard industry practice would dictate that any special equipment or test apparatus that is used to make a critical airworthiness decision or that requires calibration or inspection be given a unique part number and serial number to identify it within the certificate holder's inventory system.

C. Ensure that the OEM maintains the records dealing with special equipment or test apparatus in a manner acceptable to the FAA for as long as the records are used for making airworthiness determinations.

6-2648: Task outcomes

A. Complete the PTRS.

B. File all supporting paperwork in the operator's/applicant's office file.

6-2649: Future activities

Conduct normal surveillance of the repair station's IPM or air carrier's CAMP to ensure that procedures are adequate.

Paragraphs 6-2650 through 6-2665 are reserved. Uncontrolled copy when downloaded. Check FSIMS (http://fsims.faa.gov) to verify this version is current prior to use.

Appendix B
A Complex Risk Management Program

The purpose of this document is to provide guidelines that allow for the creation of a consistent and documented method of performing risk management within Boogaloo Bearings Inc. (BBI).

SCOPE

The guidelines in this document apply to risk management analysis for products under consideration for customers of BBI.

DEFINITIONS

- *Candidate risk:* A concern that is identified during the product realization phase of production at BBI.
- *Consequence:* The quantitatively or qualitatively expressed outcome of a future occurrence, such as a loss, injury, disadvantage, or gain.
- *Consequence category:* A functional area in which a risk can affect a project. Consequence categories used in this document are safety, performance, cost, and schedule.
- *Contingency plan:* A response plan designed to limit the consequence of risk realization when the primary mitigating actions prove ineffective.
- *Impact horizon:* Allows for the categorization of impact time frames in relation to the current date. It represents an abstract time frame in which the risk may occur. Impact horizon values can be near, mid, or long term.
- *Impact time frame:* Represents the time when the risk may occur. It consists of two pieces of data: a sunrise date, which indicates the earliest time the risk could occur, and a sunset date, which indicates the latest time the risk could occur.
- *Issue:* A risk that has occurred and has become a problem. It is an undesirable event whose occurrence cannot be stopped or directly controlled.
- *Likelihood:* The probability of a specified outcome.
- *Priority score:* A score that is numerically represented by a cross-reference of the likelihood and consequence scores of a risk plotted on a risk matrix.

- *Problem:* An existing adverse situation. A problem is a risk that has already occurred or is certain to occur. Reactive management (not risk management) is necessary to deal with a problem because a problem can lead the project into new risks.
- *Risk:* A measure of the potential inability to achieve a goal or target within defined safety, cost, schedule, and technical constraints. A risk has two components: the likelihood of failing to achieve a particular outcome and the consequence of failing to achieve that outcome. A risk is a future event and can often be managed proactively.
- *Risk acceptance:* A risk can be accepted when its consequences are tolerable or can be reasonably mitigated.
- *Risk analysis:* Examines risks in detail to determine their extent and relationships. Risk analysis also classifies risks and ranks them according to importance, and it evaluates all identified risks to estimate the likelihood of occurrence, consequence of occurrence, and time frame for necessary mitigation actions.
- *Risk approval:* The decision to validate a candidate risk. Risk approval can be performed by BBI project managers or executives.
- *Risk assessment:* The qualitative and/or quantitative evaluation of the likelihood and consequence of a risk occurring.
- *Risk attribute:* A characteristic of likelihood and consequence that describes or defines standard ways of assessing the consequence or success of a risk mitigation plan. Risk attributes are chosen during risk planning and provide meaningful information that can enable more informed control decisions.
- *Risk classification:* Includes the processes of:
  - —Grouping risks into high, moderate, or low categories based on the likelihood and consequence adjective ratings
  - —Grouping risks based on shared characteristics or relationships
- *Risk closure:* A risk that either no longer exists (i.e., the risk has evolved into a problem or the sunset date has passed), is no longer cost-effective to track, or has been mitigated.
- *Risk escalation:* The process of raising attention of risks to BBI's top management.
- *Risk identification:* Examines each element of a project to identify and document risks that may affect BBI. Risk identification begins as early as possible in a successful project and continues throughout its life cycle.
- *Risk management:* A process that encompasses identification, analysis, mitigation planning, and tracking of root causes and their consequences.
- *Risk management planning:* Develops and documents an organized, comprehensive, and interactive strategy for identifying and tracking root causes, developing risk-handling plans, performing continuous risk assessments, and assigning adequate resources.

- *Risk matrix:* A graphical representation of a risk's likelihood and consequence scores. Sometimes called a "5 × 5 matrix," it has five rows and five columns. The rows show likelihood scores and the columns show consequence scores. Each cell can be represented by a priority score.
- *Risk mitigation:* Lessens the severity of a risk by reducing the likelihood of its occurrence and/or minimizing the consequences of occurrence.
- *Risk mitigation plan:* A document that captures the actions to be taken to reduce the likelihood of risk occurrence. This document is the output of risk mitigation planning.
- *Risk mitigation planning:* The process of analyzing a risk to determine actions that may be taken to reduce the likelihood of risk occurrence.
- *Risk owner:* An individual to whom a risk is assigned. This person is responsible and accountable for identifying, implementing, and tracking the risk mitigation approach and actions.
- *Risk research:* The investigation of an identified risk. Risk research continues until there is enough information to determine whether risk ownership is still properly assigned and until there is a risk-handling strategy.
- *Risk review:* The objective of the risk review is to provide a venue for the risk team to communicate to BBI management any risks that may affect the company. These reviews are intended to inform and elicit feedback from management regarding the risks and associated handling plans.
- *Risk team:* The objective of a risk team is to understand and approve risks, evaluate them against the costs to mitigate them, and direct/approve the risk-handling approaches and risk mitigation activities.
- *Risk team member:* Personnel assigned to work on a proposed product review as part of a quality management system review of proposed projects. Risk team members are responsible for bringing potential risks to the attention of their program managers. Risk team members may also collect data to assist in the monitoring and tracking of a risk.
- *Risk tracking:* Captures, compiles, and reports risk attributes and metrics that determine whether risks are being mitigated effectively and whether risk mitigation plans are implemented correctly.

ACRONYMS

BBIPR: BBI procedural requirement

ECD: estimated completion date

FOCR: functional organization council review

FY: fiscal year

OSHA: Occupational Safety and Health Administration

PDR: preliminary design review

PM: program manager

QM: quality manual

RMS: risk management system

IDENTIFYING RISKS

All risks and associated information shall be determined for each program under development. The program manager (PM) shall develop an implementation plan for conducting risk management. The basic risk management strategy is intended to identify both technical and nontechnical critical areas and risk events, and take necessary actions to handle them before they can become problems and cause serious cost, schedule, or performance consequences.

Risk management uses a structured assessment approach to identify and analyze processes and products that are critical in meeting BBI's objectives. The PM will develop risk-handling options to mitigate the risks and will monitor their effectiveness. The identification of the resources required to implement the developed risk-handling options is key to the success of the risk management effort.

Risk information will be included in all project risk reviews. As new information becomes available, the PM will conduct additional reviews to ascertain whether new risks exist. The goal is to continually look for future risks in areas that may severely affect BBI.

As part of his or her management activities, a PM (or other team member) will identify internal risks associated with his or her project. Team members will identify external risks.

Risks can be identified using a number of processes. This document does not define how any specific risk identification process is to be performed, as different types of projects may require different approaches. However, some of the more common approaches are listed in Table B.1.

Risk identification depends heavily on both open communication and a forward-looking perspective. This encourages all personnel to communicate new risks and to plan beyond their immediate problems. Although individual contributions play a role in risk management, teamwork improves the chances of identifying new risks by allowing all personnel to combine their knowledge and understanding of the project.

Table B.1 Risk identification approaches.

System safety assessments: fault tree analysis, hazard analysis, failure mode and effects analysis	Brainstorming
Quantitative risk assessment	Test and verification
System and software engineering	Pause and learn sessions
Program planning and control: cost and schedule risk analysis	Experience: previous analysis, lessons learned, historical data
Models and simulations	

Risk identification should begin as early as possible. The PM (or a designee) shall document each risk identified to include a title, risk statement, context statement, and closure criteria, as shown in Table B.2. This is performed for both internal and external risks.

The title of the assessment should be written to attract attention and focus on the appropriate audience. It should answer the question "So what?"

The risk statement should generally be written in a condition-consequence format. For instance, given a condition, there is a possibility that a consequence will occur. The risk statement should not be equivalent to the solution; thus, do not write a risk statement that provides a solution.

The context statement provides background on the risk and should include only facts (not assumptions). Ensure that no new risks are introduced here. A well-written context statement can serve as the starting point for the development of a risk mitigation plan.

The closure criteria should document how the risk can be eliminated or reduced to an acceptable level. It should be specific and measurable. Closure is the goal of risk mitigation and is the last step in the risk mitigation plan.

The PM shall analyze and prioritize identified risks according to likelihood of occurrence, consequence, effect time frame, and impact horizon.

Risk likelihood can be determined in a quantitative or qualitative manner. In either case, the results of the analysis can be used to assign each risk a score from one to five, as illustrated in Table B.3.

The consequence score is determined by assessing the consequence of the risk and assigning a score from one to five based on the criteria in Table B.4. Risks must be analyzed and scored on each separate consequence category.

Table B.2 Risk documentation.

	Lack of titanium for use in Acme project
Risk statement	Given that there is a 23-week lead time in acquiring titanium, there is a possibility that BBI might not be able to meet the projected delivery time
Context statement	Without a solid and captured source of material, BBI might not be able to meet the March 23, 2014, delivery requirement
Closure criteria	Complete control and schedule for delivery of titanium

Table B.3 Rating risks.

Score	Likelihood of occurrence (p)	Consequences
5	Near certainty	$p > 80\%$
4	Highly likely	$60\% < p \leq 80\%$
3	Likely	$40\% < p \leq 60\%$
2	Somewhat likely	$20\% < p \leq 40\%$
1	Not likely	$p \leq 20\%$

Table B.4 Risk criteria.

	1	2	3	4	5
Performance	Minimal consequence to objectives/goals	Minor consequence to objectives/goals	Unable to achieve a particular objective/goal, but remaining goals represent better than minimum success or outcome	Unable to achieve multiple objectives/goals, but minimum success can still be achieved or claimed	Unable to achieve objectives/goals such that minimum success cannot be achieved or claimed
Safety/human	Discomfort or nuance	First-aid event per OSHA criteria	Injury or illness does not result in lost time per OSHA criteria	Injury or illness results in some lost time per OSHA criteria	Loss of life
Asset	Minimal consequence: asset has no sign of physical damage	Minor consequence: asset has cosmetic damage and is repairable	Minor consequence: asset is damaged but repairable	Major consequence: asset is substantially damaged but repairable	Destroyed: asset is compromised and un-repairable; a total loss
Schedule	Minimal consequence	Critical path is not slipped; total slack of slipped tasks will not affect critical path within 10 days	Critical path is not slipped; total slack of slipped tasks is within 10 days of affecting the critical path	Critical path slips	Critical path slips and one or more critical milestones or events cannot be met
Cost	Minimal consequence	Minor cost consequence: cost variance <5% of total approved FY baseline	Cost consequence: cost variance >5% but ≤10% of total approved FY baseline	Cost consequence: cost variance >10% but ≤15% of total approved FY baseline	Major cost consequence: cost variance >15% of total approved FY baseline

The highest score from the consequence categories is then used as the final consequence score. For instance, a risk with a likelihood score of three that has a safety consequence score of two, a performance consequence score of three, a schedule consequence score of four, and a cost consequence score of three would be considered a "3 × 4" risk. If one of the consequence categories is not applicable, its score is zero (i.e., if a risk has no consequence for safety, the safety consequence score should be zero).

In addition to likelihood and consequence scores, the impact time frame and impact horizon are determined. The impact time frame consists of two dates: sunrise and sunset. The sunrise date is the earliest time that the risk could occur, and the sunset date is the latest time that the risk could occur (i.e., when the risk can be retired). These values can be used to prioritize the risks in addition to the consequence and likelihood.

The impact horizon is determined from the dates in the impact time frame shown in Table B.5. This can be used to help further prioritize risks according to the time frame in which they will occur.

Risk scoring is performed to facilitate risk review, integration, roll-up, prioritization, and summarization. Risks should be analyzed at a level that's sufficient to determine the relative importance for planning cost-effective mitigation strategies and for tracking. For example, high-likelihood, high-severity risks require a more detailed level of analysis in order to plan effective mitigation strategies.

External risks are analyzed in the same manner described for internal risks to determine likelihood of occurrence, consequence, impact time frame, and impact horizon. However, the likelihood and consequence criteria from the development project are used in place of the criteria depicted in this document. Additionally, development projects may not use the concepts of sunrise, sunset, and impact horizon, but these values are still calculated and used in risk tracking and monitoring by project team members.

Prioritizing Risks

When the likelihood score, consequence score, impact time frame, and impact horizon have been determined, a risk can be prioritized by determining its priority score. The priority score is the score assigned via the risk matrix cell into which the risk falls, and it is determined from the risk's likelihood and consequence scores (see Table B.6). For example, a risk with a likelihood score of 3 and a consequence score of 4 would yield a priority score of 19.

Development projects may or may not have prioritization processes or criteria. If the development project has such processes, follow them to prioritize external

Table B.5 Time frames.

Impact	Sunrise	Sunset
Near	<2 months	<2 months or 2–6 months
Mid	2–6 months	2–6 months or >6 months
Long	>6 months	>6 months

Table B.6 Risk likelihood and consequences.

		Consequence				
		1	2	3	4	5
Likelihood	5	7	15	20	23	25
	4	6	13	18	22	24
	3	4	11	15	19	21
	2	2	8	10	14	17
	1	1	8	5	9	12

risks. If the development project does not have a prioritization process, use the priority score to determine the level of review required for the risk.

Risk prioritization is part of risk analysis and provides a basis for allocating handling and contingency resources, developing handling plans, and preparing individual handling tasks. A key objective in prioritizing risks is to determine which risks will be fully accepted, actively controlled, closely tracked, or only watched. The risk prioritization process is closely related to the escalation process.

Risk Review

Once the PM has completed the analysis of the risks, the risks are submitted to the risk management team so it can assess them for completeness and consistency. The primary intent of this review is to ensure that the risks are adequately documented to include the information shown in Table B.7 and that the scoring is indicative of the risk statement.

At this point, the process diverges into two distinct processes. If the risk is internal, the process continues with section 4.4, Planning. If the risk is external, the process continues with section 4.5, Communicate, control, and track risks. The risk review process is performed at various points throughout a risk's life cycle. The specific review performed is dependent on the maturity and priority of the risk.

The risk management team, PM, and top management perform risk reviews. In all cases, the goal is to assess risks for completeness and consistency and provide initial communication of the risks to BBI's management. In all risk reviews, the risk is returned to the PM or risk owner for corrective action if the risk does not receive concurrence.

Planning

Planning decides what should be done about a risk. In this process, decisions and handling strategies are developed based on current knowledge of project risks. The planning process applies only to internal risks.

When the initial risk review is complete, the risk enters a planning phase. During this phase, the risk is further analyzed to ensure that the consequence and likelihood scores are correct and that the risk has the correct owner. Minimally, each risk attribute (leading indicator) shall be identified for each risk. The risk attributes can

Table B.7 Risk management team review.

Maturity	Priority	Risk management team review	Functional lead risk board review	Top management risk review
Risk (only)	<8	X	X	
	>7 and <17	X	X	
	>16	X	X	X
Risk mitigation plan	<8	N/A	N/A	N/A
	>7 and <17	X	X	
	>16	X	X	X

Table B.8 Risk follow-up.

	Priority <8	Priority >7 and <17	Priority >16
Research	Determine risk attributes for tracking. No other analysis required.	Determine risk attributes for tracking. Perform research to increase confidence in scoring and to determine mitigation approaches.	Determine risk attributes for tracking. Perform research to increase confidence in scoring and to determine mitigation approaches and contingency options.
Plan	Document risk attributes being used for tracking.	Document mitigation approach with milestones to show likelihood reductions. Include risk attributes that will be used for trending.	Document mitigation approach with milestones to show likelihood reductions. Document contingency plans. Include risk attributes that will be used for trending.

be used to track the status of the risk (trending). Beyond minimum requirements, the PM or the risk owner shall develop plans for each risk, as shown in Table B.8.

The research phase concludes with either risk ownership affirmation or risk reassignment and the development of a plan that recommends one of two risk approach strategies: "watch" or "mitigate."

Watch

Risks can be watched if one of the following conditions exists:

- The risk is low priority (priority less than eight)
- Sufficient mitigation resources are not available

- Mitigation cost is comparable to recovery cost if the risk were to occur
- The likelihood of mitigation success is uncertain

Watched risks can be placed on a list and periodically (at least quarterly) reassessed. Their risk attributes are continually monitored for early warning of critical changes in likelihood, consequence, time frame, or other aspects.

To watch a risk, the PM or the risk owner will create a metric to indicate a trend in the priority score or specific trigger points/dates when the risk is to be assessed for specific conditions or attributes. When the metric is exceeded, metric thresholds are set to trigger the implementation of specific risk mitigation plans or a reevaluation of the risk. Deviation of the metric from predetermined criteria may also be an appropriate risk trend indicator.

Mitigate

Risk mitigation reduces the likelihood of risk occurrence or shifts the time frame in which the risk will occur through the execution of a predetermined set of action steps (e.g., risk mitigation plan). The execution of a risk mitigation plan may require additional resources; however, the selected approach may result in an acceptable risk level.

It is the goal of BBI to minimize risks to the lowest practical level within the allocated resources. When choosing to mitigate a risk, the following questions should be considered:

- Cost
  - —Is the risk mitigation plan within the current budget and scope?
  - —How much does each mitigating option cost? How likely is it to succeed?
  - —Will the mitigation cost more than the actual cost of the risk consequence?
- Schedule
  - —Is each mitigation task part of the in-scope work?
  - —Which tasks are new work that should be addressed through the BBI budget processes?
  - —Who is responsible for each of the mitigating tasks?
  - —What is the consequence of the project schedule for each mitigation option?
  - —Does the risk impact a critical or significant milestone? Are there clear references to these consequences?
  - —Are all stakeholders identified and included in the mitigation planning?
- What is the confidence level for successful completion of each mitigation option?

- How much will risk be reduced at the completion of the risk mitigation plan?
- What are the costs of mitigation versus the benefits and uncertainties of risk reduction?

All risk mitigation plans are reviewed and concurred upon at the appropriate escalation level to determine which risk mitigation activities will be actively pursued. (See section 4.5, Communicate, control, and track risks.) Ownership of these risk mitigation plans is assigned to the appropriate level for execution. If the choice is to mitigate the risk, the risk owner develops a detailed risk mitigation plan. Each plan may have several action steps that need to be performed for effective mitigation. The risk owner implements the plan and oversees completion of all its actions until the risk reaches an acceptable limit or is eliminated. At that time, the risk owner recommends to the appropriate management that the risk is accepted (residual risk remaining) or closed (risk eliminated). Depending on the amount of residual risk remaining, the risk can either remain open or move into the "watch" state.

Accepted mitigation activities are integrated into the project schedule to ensure that the schedule accounts for all task dependencies and to monitor progress on the risk management plan. In some instances, a mitigation task may represent new work that requires a baseline revision (or other approval).

Risk mitigation plans are approved and controlled by the PM.

Contingency Planning

Risks with priority scores greater than 16 also require a contingency plan. The contingency plan applies in the event that risk mitigation efforts fail. This allows BBI to proactively plan for potential problems.

The risk owner develops the contingency plan and links it to the appropriate risk record. It must be approved by top management. The development of a contingency plan may be waived with approval from the PM.

Risk-Handling Plan Review

Any internal risk with a priority score greater than seven requires the PM to discuss the risk-handling plan with the risk management team. This step will ensure quality assurance and concurrence on the risk-handling plan and that contingency actions will be taken if the risk becomes a problem. The review is not an approval of the risk-handling plan; rather, it's an assessment of its completeness and consistency.

Communicating the risk-handling plan to the risk management team is valuable in determining whether there may be other potential risk stakeholders, ensuring that requests for risk transfers are acted on, and raising significant concerns with the PM.

There is no requirement to generate a risk mitigation or contingency plan for external risks. BBI isn't responsible for the mitigation of external risks. The development project is responsible for mitigating its own risks. However, in some instances BBI may wish to develop a suggested mitigation approach for the development project.

Internal risks with priority scores of less than eight don't require risk mitigation plans. However, all other risks require one. Risk mitigation and contingency plans are developed by the PM. If the plans don't gain concurrence from the risk management team or the functional lead risk board, they should be returned to the risk owner for corrective action.

COMMUNICATE, CONTROL, AND TRACK RISKS

Risk communication depends on risk escalation. Risk escalation is the formal communication of internal risks throughout BBI and the formal communication of external risks with development projects.

Risk escalation results in risk prioritization, proper risk integration, and risk transfer. It's aided by the risk management team and through risk tracking and control. Each risk escalation level evaluates, controls, and oversees those risks that require that level's attention. Any risk has the potential for escalation. A risk may be escalated for several reasons. In general, they fall into the following four categories:

- Management awareness or its decision is needed due to the potential severity or consequences of the risk (BBI has established criteria for assessing the need for escalation-based risk priority)
- Additional resources requested for effective mitigation
- Potential transfer of the risk to another organization or level (i.e., a risk identified at the project level is actually a programmatic risk, and ownership should be transferred accordingly)
- Coordination, integration, or transfer is needed with other organizations or stakeholders internal or external to BBI

In addition, any individual risk can be escalated to multiple levels simultaneously.

Internal Risks

After they're identified and analyzed, internal risks are communicated to the PM prior to the execution of any planning, as noted in section 4.4, Planning. The goal of this communication is to make management aware of any new risks that have been documented prior to the planning stage.

Once risk planning is complete, the plan is approved by the PM. This only applies to risks categorized as yellow or red (see section 4.4, Planning). The goal of this communication is to make top management aware of the planned approach for dealing with risks and to provide concurrence on the risk handling plan. This communication path also provides the status of any ongoing risks on each department's top risk list.

Risk Reviews

The core objective of risk reviews is to make top management aware of the planned approaches for dealing with risks and to provide information on risk handling plans.

Reviews provide overviews of existing risks. The overview can be as simple as a list of risks with priority scores less than 17 with their locations on the risk matrix.

Risks with priority scores greater than 17 need to be reviewed in detail.

Control

Risk control includes the evaluation of risk tracking reports. Risk mitigation tracking reports are evaluated for adequate progress (actual vs. planned) and for validation that appropriate corrective actions have been implemented. The following strategies can be selected during risk control:

- Continue as planned
- Invoke a contingency plan (if the risk has become a problem)
- Replan (update or develop a new risk mitigation plan)
- Close or accept the risk

All decisions and directions regarding risks and risk mitigation plans should be documented and maintained.

Risk Closure

To close a risk, the PM must concur with the closure rationale. If the residual risk has a priority greater than five, it cannot be closed (the risk must either be further mitigated or be accepted). Any risk with a priority of five or less is assumed to be sufficiently mitigated and may be closed. A risk may also be closed if it has evolved into a problem or if its sunset date has passed.

Risks can be accepted during risk control if mitigation efforts don't result in a priority score that's less than five and it's determined that further mitigation isn't practical.

Risk Acceptance

If a risk is accepted, it doesn't require further action other than periodic monitoring. A risk can be accepted when its consequences are tolerable or when it cannot be cost-effectively mitigated with further action. Generally, if risks aren't fully mitigated (having a priority score greater than five) and there are no plans for further mitigation, they can be considered accepted unless the selected handling strategy is "watch." Accepted risks must be reviewed periodically (at least quarterly) to ensure that conditions and assumptions about them haven't changed. Accepted risks could potentially be reopened or closed on the basis of these reviews. Written approval from top management is required before a risk can be accepted, but the ultimate decision is left up to the president of BBI.

This acceptance must be documented and provide the reasons and conditions under which the risk was accepted. It's important that the rationale be clear because these are important factors considered during the periodic reviews.

Tracking

The PM shall track the status of the risk and the effectiveness of risk handling actions against established metrics (previously identified risk attributes). The objective of risk tracking is to collect accurate, timely, and relevant risk information to present to the appropriate decision maker. Risk tracking is performed to measure actual versus planned progress associated with implemented risk mitigation plans. Risk tracking also provides input into the risk escalation process.

Each risk should already have an attribute/indicator system designed to provide early warning of changes in status. This allows management to respond proactively before the risk becomes a problem.

Risk tracking isn't a problem-solving technique. Rather, it's a proactive technique to provide a basis for developing additional risk handling options and/or approaches, updating existing risk handling strategies, and/or reanalyzing known risks. In some cases, tracking results may also be used to identify new risks and revise some aspects of risk planning.

Risks can change over time. Every action taken or not taken changes the basic facts from which each risk is derived. Continually monitoring risks and reassessing their potential consequence (i.e., repeating the risk identification, assessment, and handling steps at periodic intervals) is essential for appropriately managing risks.

Appendix C
A Simple Risk Management Program

When considering potential new large business, Boogaloo Bearings Incorporated completes a risk assessment checklist. The purpose of this assessment is to ensure that the company does not take on a new program or business that could potentially hurt its performance on existing business and/or new business. If the new program is documented as being high risk after the checklist is complete, the president of Boogaloo will need to sign off on the program before the quote is submitted to the customer. At that point Boogaloo understands that if it wins the bid, it will be at risk and all steps to mitigate the risk must be followed. A high-risk assessment score from the checklist may result in Boogaloo sending a "no bid" and a thank-you to the customer if Boogaloo decides that it does not want to accept the risk.

6.1 Assignment of responsibilities of risk management—It is the duty of the estimating department personnel to begin the risk assessment of any potential new large business. This is part of the quoting process. The estimating department, along with input from the president, sales department, and quality department, will begin the process using a checklist.

6.2 Definition of risk criteria—Boogaloo has defined the level at which a new large program shall be required to undergo a risk assessment to be any program with potential annual sales of over $1 million. This includes new business or a change to an existing contract that has a delta of over $1 million a year. If a program is under this requirement, it does not have to have an assessment completed, but could have one completed at the discretion of anyone involved.

6.3 Identification, assessment, and communication of risks throughout the product realization—The estimating department will use the risk assessment checklist (see Figure C.1) to help communicate the risks for that particular job. All information related to the job will be documented on the form and kept in the job folder. Use the checklist to identify the ranking of the risk itself as it pertains to technical, schedule, or cost issues with the program. These rankings should be filled out using the checklist.

6.4 Identification, implementation, and management of actions to mitigate risks that exceed the defined risk acceptance criteria—The completed form will be reviewed to ensure that all risks have been identified. The

Quote number: ____________	Date: ____________
Program name: ____________	Customer: ____________
Annual value: ____________	Estimator: ____________

List potential risk	**Score (1–5)**
1	
2	
3	
4	

Risk mitigation		
Process/responsible	Risk	Mitigation
1		
2		
3		
4		
Approval: ____________	Date: ____________	

Figure C.1 Risk assessment form.

final risk score will be analyzed, and steps to mitigate the risk will be documented.

6.5 Acceptance of risks remaining after implementation of all mitigating actions—After the form has been completed and all actions to mitigate risks have been listed, the president will sign off on the form. This will ensure that all risks were understood at the time of acceptance of submitting a quote for the new large program.

6.6 Mitigation will be performed to eliminate or minimize the probability of a risk's occurrence or the impact of the risk should it occur throughout product realization. These risks could include, but are not limited to, material availability, nonconforming product, lack of approved suppliers, lack of trained/qualified personnel, and so on. Risk mitigation will be documented on the risk mitigation form.

Checklist Instructions

1. Fill in the quote number, date, program name, customer, and program annual value. The program annual value must be greater than $25,000 to use this form. If it is less than $25,000, this form can be used on an as-needed basis.

2. Fill in all potential risk for this program and rank them from one to five, with one being the smallest risk and five being the greatest risk. Additional risks can be added more than five, if necessary. An example of extra added risk is as follows: The new program requires equipment that company does not currently own. This will be a large risk to purchase the equipment and set it up for use (five points of risk). The titanium suppliers have a 52-week lead time, but the titanium materials are not available for a six-month delivery (five points of risk). This example is a risk score of 10.
3. Analyze the scores. Any individual score of two or less does not need to have any of the risks mitigated. This program would not be considered a risk to the total performance of the company. An individual score of three or higher will need to have documented mitigations and will also need to have the president's signature to submit the bid to the customer. An example of a documented mitigation is the following: Equipment will be procured from the customer and will move with the job, along with the employee who has been using the equipment for the past five years.
4. Complete form as required and file it in the quote folder if signed.

Risk number				
1	2	3	4	5
Very unlikely to occur; strong controls in place	Not likely to occur; controls have minor limitations/ uncertainties	May occur; controls exist with some uncertainties	Highly likely to occur; controls have significant uncertainties	Nearly certain to occur; controls have little or no effect

Appendix D
Sample Fluorescent Penetrant Inspection Procedure

1.0: Purpose

1.1 The purpose of this procedure is to define the minimum requirements for the control and performance of fluorescent penetrant inspection (FPI) performed at Boogaloo Aircraft Repair Inc. (BARI).

2.0: Scope

2.1 This document does not apply to personnel qualified and certified to perform FPI per OP 17.1.

2.2 This procedure is mandatory and applies to all products inspected by the FPI method, including reworked parts. This document meets or exceeds the requirements of MIL-STD-6866, ASTM E1417, and all applicable customer specifications. In the event of a conflict between this document and customer requirements, the customer requirements shall have precedence.

2.3 Customer products having specialized requirements will be detailed as appendices to this procedure. The applicable appendix shall be referred to for any processing or inspection parameters that differ from this procedure. In such cases, the customer requirements spelled out in the applicable appendix will have precedence.

3.0: Referenced documents

3.1 The following documents have been referenced in the preparation of this procedure and are considered part of this procedure as applicable. (*Note: These relate to AS9100's maintenance documentation requirements.*)

— ASTM E 1417: Standard practice for liquid examination penetrants

— AMS 2440: Inspection of ground and chromium-plated steel parts

— MIL-STD-6866: Penetrant inspection method

— MIL-STD-410: Qualification of inspection personnel

— QPL25135: Military qualified products list

— QPL-AMS-2644: Qualified product list, inspection material, penetrants

— MIL-I-25135E: Inspection materials, penetrants

— SNT-TC-1A: Personnel qualification and certification in nondestructive testing (NDT)

— NAS 410 NAS: Certification and qualification of NDT personnel

— OP 0017.1: Qualification of NDT personnel

3.2 Customer documents are considered part of this procedure as applicable.

4.0: Responsibility

4.1 The manager of quality assurance directs the administration of this procedure and is responsible for the certification of the Boogaloo aircraft repair penetrant level III. The level III is responsible for the implementation of this procedure and for proper training and certification of FPI inspection personnel.

4.2 The quality assurance manager and the level III consultant will review this procedure biannually to ensure that all industry standards are met. Evidence of this review will be documented in the procedure review log sheet. (*Note: This relates to AS9100's record-keeping requirements.*)

5.0: Personnel

5.1 Personnel performing FPI to this procedure shall be qualified and certified in accordance with OP 0017.1.

5.2 A level I shall be able to process parts for FPI under the supervision of a certified FPI level II or III. This includes penetrant application, removal, drying, and developing. Only a certified FPI level II or III shall accept or reject product for final disposition. (*Note: This relates to AS9100's training and training record requirements.*)

6.0: Materials

6.1 Unless otherwise directed by customer requirements, and if approved by the level III inspector, only materials listed in QPL-AMS-2644 shall be used.

6.2 The penetrant used shall be fluorescent, water washable (Type I, Method A), sensitivity level 3: Sherwin HM 604.

6.3 The developer used shall be a dry developer (Form A): Met-L-Chek D72A.

6.4 A quick-drying solvent, Met-L-Chek R-504, shall be used to remove excess penetrant when evaluating indications.

6.5 A nonaqueous wet developer, Sherwin D100, shall be used when required by the customer or when evaluating indications. (*Note: This relates to the standard's purchasing requirements.*)

6.6 The water used for removing excess penetrant shall be free of contaminants that would leave a residue on the component when dried.

7.0: Equipment

7.1 The penetrant inspection equipment shall be maintained to permit uniform, controlled penetrant application and efficient removal, part drying, developer application, and part evaluation. The equipment shall be adequately protected when not in use in order to prevent contamination of materials. Timers, temperature gages, and water and air pressure regulators shall be provided for necessary controls.

7.2 *Penetrant application tank:* Penetrant may be applied to the part by dipping, brushing, or spraying or by pouring it over the part.

7.3 *Rinse station:* The rinse station must be supplied with temperature- and pressure-controlled water. The excess penetrant must be removed with a Tri-Con 400501 spray nozzle or its equivalent. An overhead black light is used to ensure that adequate washing is performed. (*Note: This relates to the standard's purchasing requirements.*)

7.4 *Drying oven:* A recirculating hot-air drying oven is used to dry the part after the excess water has been removed. The oven is equipped with a timer and temperature controls.

7.5 *Developer tank:* The dry developer is applied by placing the part in the developer tank and pressing the start pedal or by dusting the developer over the part.

7.6 *Inspection booth:* Parts are evaluated in a dark booth using black light(s) operating in the range of 320–380 nm. The booth is equipped with white lights, mirrors, magnifiers, and so on, to aid in proper evaluation.

7.7 The penetrant area shall be checked daily for cleanliness and excess fluorescence.

8.0: Precleaning

8.1 All surfaces to be inspected shall be clean, dry, and free of contaminants or foreign materials that could prevent the penetrant from entering discontinuities, suppress dye performance, or produce unacceptable background indications.

8.2 Components will be precleaned by aqueous degreasing unless other cleaning methods are dictated by surface condition or customer requirements. An etch operation shall be required if any metal smearing operation has been performed on the component being inspected. Components that require pre-penetrant etch shall be sent to an approved outside vendor for the etch operation.

8.3 When aqueous cleaning solutions are used to prepare parts for penetrant inspection, the parts shall be oven dried at a temperature of 200°F ± 25° for 30 minutes prior to inspection. If parts are not oven dried, air drying for one to two hours shall be required with a visual inspection verifying parts are dry. Ensure part temperature is below 100°F before applying the penetrant.

8.4 A final penetrant inspection shall be performed after the completion of all operations that could cause surface discontinuities, or as dictated by customer requirements.

9.0: Penetrant application

9.1 The ambient temperature, penetrant material, and parts to be inspected shall be in the temperature range of 50°F–100°F.

9.2 Penetrant may be applied by dipping, spraying, or brushing or by pouring it over the part. It should completely cover the surfaces to be inspected. Verify proper coverage using ultraviolet light. Parts shall not be immersed in penetrant for more than half of the total dwell time.

9.3 After the penetrant is applied, parts shall be positioned to allow the oil to drain and to avoid pooling in recesses. Parts shall be repositioned as necessary during the dwell time to allow for proper draining.

9.4 Parts must be completely reprocessed if penetrant dries on the part surface.

9.5 Dwell time shall be between 10 and 60 minutes, unless otherwise specified. If ambient temperature is between 40°F and 50°F, dwell time shall be doubled. Maximum dwell time includes removal time.

10.0: Excess penetrant removal

10.1 After the appropriate dwell time, excess penetrant shall be removed in the darkened rinse area. A minimum black light intensity of 100 $\mu W/cm^2$ at the part surface is required, and a maximum of 10 foot-candles of ambient white light is allowed when removing the excess penetrant. Avoid overwashing by monitoring the fluorescent background on the part.

10.2 Rinse-water temperature shall be maintained between 50°F and 90°F. The water pressure shall not exceed 30 psi. The water nozzle shall be positioned at least 12 inches from the surface being washed, and held at an angle of approximately 45° to the part surface. Large or complex parts shall be washed from the bottom up.

10.3 Rotate part to remove all traces of excess water. Clean, filtered air at a maximum of 25 psi may be used to assist in excess water removal. The air gun nozzle shall be held at least 12 inches from part surface.

11.0: Dry

11.1 Place parts in the drying oven in a position where the water can properly drain. Reposition parts during the drying cycle as required. The oven temperature shall be maintained between 110°F and 140°F. Parts shall be removed from the oven upon drying, but in no case remain in the oven longer than 20 minutes.

11.2 Pressurized air shall not be used to help dry the part or to remove water after the part has been removed from the dryer.

12.0: Develop

12.1 Parts shall be allowed to cool to 110°F or below prior to application of developer. The use of cooling fans is allowed. Developer must be applied within 30 minutes of the completed drying cycle.

12.2 Place dry part in dry developer tank or spray on a light, even coat of nonaqueous developer. Ensure that developer lightly and evenly covers 100% of the area of interest. Dwell time is 10 minutes to 1 hour for nonaqueous developer and 10 minutes to 4 hours for dry developer, unless otherwise specified in customer requirements. As an option, the dry developer may be dusted over the area of interest.

12.3 After the appropriate dwell time has elapsed, the excess developer may be removed by lightly tapping the part or by using clean, filtered air at a maximum of 5 psi.

13.0: Inspection

13.1 Inspectors shall not wear sunglasses or eyeglasses with photochromatic lenses while inspecting parts.

13.2 Inspectors shall wait a minimum of one minute before beginning or continuing inspection in the event that any of the following occur:

— White light is turned on

— Inspector looks directly into the black light

— Inspector enters viewing area from ordinary room light

13.3 Allow a minimum of 15 minutes for the black light to reach full intensity. The minimum black light intensity shall be 1200 $\mu W/cm^2$ when measured at the part surface or at 15 inches from the black light filter. Parts must be evaluated within 15 inches of the filter, or the intensity at the part surface must be verified to be at least 1200 $\mu W/cm^2$ at the distance being used.

13.4 The maximum ambient white light at the part surface shall not exceed 20 lux or 2 foot-candles. A white light source capable of emitting a minimum of 100 foot-candles at the inspection surface shall be available to assist in the evaluation of discontinuities.

13.5 Parts shall be inspected before the maximum developer dwell time has elapsed. Parts not inspected during this time must be cleaned and reprocessed. If the interval between excess penetrant removal and inspection exceeds three hours, the part must be reprocessed.

13.6 Acceptance criteria shall be as specified in the applicable purchase order, drawings, blueprint, ordering data, or shop work order. When no acceptance criterion is specified, the customer shall be contacted to provide the information. Unless otherwise specified, 100% inspection is required. (*Note: This relates to the standard's maintenance documentation requirements.*)

13.7 Parts that exhibit excessive background fluorescence or appear to be overwashed shall be completely reprocessed.

14.0: Reevaluation of indications

Note: Reevaluation may be performed only once unless otherwise specified by the customer.

14.1 Indications that are not readily interpretable must be reevaluated using the following technique:

— Lightly moisten a cotton swab with Met-L-Chek 504 solvent or equivalent.

— Wipe the indication once while rolling the swab, removing the penetrant bleed-out. After the solvent has dried, apply a light coat of nonaqueous or dry developer.

— Begin evaluation immediately to observe size, shape, and quickness of bleed-out. Bake final evaluation between 10 and 60 minutes dwell time.

— The decision to accept, reject, or reprocess the part shall be made in accordance with the customer specification.

15.0: After cleaning

15.1 All penetrant materials shall be completely removed from the part after final penetrant inspection if the residues appear to be detrimental to subsequent operations or the component's intended function. Remove developer and penetrant residue with water spray and brush or aqueous degreasing as necessary.

16.0: Records

16.1 Maintain a record of all parts inspected by entering the part number, quantity inspected, the disposition, date, and inspector identification in the department logbook.

16.2 A certificate of conformance report shall be completed for each part or group of parts inspected. The information required on this form includes:

— Heat number, lot number, and lot quantity

— Part number and serial number(s), if applicable

— Specifications

— Quantity inspected, accepted, and rejected

— Material and alloy

— Date of inspection and inspector identification

(*Note: These specifications relate to the standard's record-keeping requirements.*)

17.0: Marking

17.1 Parts that have satisfactorily met the penetrant inspection acceptance criteria shall be marked as required by the customer's specification. The letter "P," along with "SV" and the inspector's ID number, denotes 100% inspection. (See OP 17.4, Inspection Stamps for NDT.) (*Note: This relates to the requirement's maintenance documentation requirements.*)

18.0: Process controls

18.1 Operators shall remain alert to any changes in performance, color, odor, consistency, or appearance of all penetrant materials. Any change noted will be brought to the attention of the level III for appropriate action.

18.2 The results of all material and equipment tests shall be documented on the process control log. Records of tests performed shall be maintained for 10 years. These records shall include log sheets, calibration records, and certifications.

18.3 Keep a minimum of one quart of penetrant and one quart of developer from the same batch that is being used as the hold-out sample. Take these samples when the new batch is first used. Identify the sealed sample with a batch number and date taken. All in-house comparison tests shall be performed using this sample. Store hold-out samples out of direct and ultraviolet light.

18.4 All penetrant materials shall be properly identified by part number, batch number, and date put into service. Retain material batch certification in the inspection area. (*Note: This relates to the standard's record-keeping requirements.*)

MAINTENANCE AND CHECKS

See Table D-1 for a checklist.

19.0: System performance/penetrant sensitivity (each shift)

19.1 At the beginning of each shift, two PSM-5 panels are to be processed through the system. Ensure that the panels are clean and dry. Check panels under black light to ensure that no fluorescent indications are visible. The panel marked N-U (new, unused) is to be processed using the new penetrant and developer from the hold-out samples. The panel marked I-U (in use) shall be processed using the penetrant and developer currently in use. The panels shall be covered with the appropriate penetrants and given a 10-minute dwell time. Rinse, and then dry the panels in the recirculating hot-air oven. Do not overdry the panels.

After the appropriate drying time, place both panels in the developer chamber and develop for 10 minutes. Compare the indications on the I-U panel with the indications on the N-U panel. The number and appearance of the indications should look essentially the same. The number of indications shall not be less than the number seen when the in-use penetrant was first used.

19.2 If the I-U test panel shows abnormal indications, a master panel shall be processed using the same parameters. If the results are less than required, the penetrant system shall be shut down until the malfunction is identified and corrected. If the master panel indicates the system is under control but the I-U panel indicates otherwise, the system shall be replaced or reconditioned.

19.3 Clean panels in aqueous degreaser and then dry and store them in acetone or MEK. (*Note: This relates to the standard's storage requirements.*)

19.4 The TAM panels shall be checked monthly for degradation. Process the panels normally and measure the diameter of the indications with a certified measuring device. Compare the dimensions with the dimensions taken when the panels were first received. The panels are acceptable if the in-use indications are within 30% of the indications when new.

19.5 The degradation check precludes the requirement to perform an annual replacement or recalibration. (*Note: Items 19.4 and 19.5 relate to the standard's calibration requirements.*)

20.0: Penetrant contamination/separation (daily)

20.1 The tests for contamination and separation consist of taking a sample of the in-use penetrant at the beginning of the day in a clear container of at least six ounces. Let the sealed sample sit undisturbed for at least four hours. Check the tank for surface scum or other evidence of contamination.

Table D.1 Process control test frequency.

Material	Frequency	Requirement
System performance (TAM panel)	Each shift	19.0
Penetrant sensitivity (TAM panel)	Each shift	19.0
Dry developer	Daily	Not lumpy or caked
Penetrant contamination	Daily	None
Penetrant separation	Daily	None
Penetrant comparison	Daily	Essentially the same
Penetrant brightness (in-house)	Daily	Essentially the same
Penetrant sensitivity	Monthly*	Four cracks minimum
Penetrant water content	Monthly*	5% max
Penetrant brightness	Monthly*	90%–110% of standard
Penetrant removability	Monthly*	Satisfactory

Equipment task	Frequency	Requirement
Rinse-water pressure	Each shift	30 psi max
Rinse-water temperature	Each shift	50°F–90°F
Drying oven temperature	Each shift	110°F–145°F
Black light filters	Each shift	Clean, undamaged
Black light (inspection booth)	Each shift	1,200 $\mu W/cm^2$ at 15 inches
Ambient light (inspection booth)	Each shift	2 foot-candles maximum
White light intensity	Each shift	100 foot-candles minimum at part
Black light (wash booth)	Each shift	100 $\mu W/cm^2$ @ part
Black light borescope (before use)	Each shift	1000 $\mu W/cm^2$
Ambient light (wash booth)	Each shift	10 foot-candles at part
Air pressure (developer removal)	Each shift	5 psi maximum
Air pressure (pre-drying oven)	Each shift	25 psi maximum
Penetrant area	Daily	Clean
Air lines cleanliness	Daily	Clean air
TAM panel degradation	Monthly	Within 30% of new
Pressure gages	Biannually*	±5 psi
Temperature gages	Biannually*	±5°
Timers/meters	Biannually*	Manufacturer specs
Drying oven	Quarterly	±10°
Penetrant tank	Annually	Clean

*By outside vendor; see approved supplier list

20.2 After four hours, any obvious separation of constituents, cloudiness, waxy deposits, unusual color, or odor is cause to replace the penetrant. (*Note: Items 20.1 and 20.2 relate to the standard's calibration requirements.*)

21.0: Penetrant comparison/in-house brightness (daily)

21.1 The comparison and in-house brightness test consists of putting a drop of the in-use penetrant and a drop of the hold-out sample penetrant side by side on a piece of light-colored filter paper or a paper towel. As the two drops merge, the color and brightness should look essentially the same when viewed under black light. (*Note: Item 21.1 relates to the standard's calibration requirements.*)

22.0: Sensitivity, water content, brightness, removability (monthly)

22.1 Monthly, a sample from the working batch that is identified by batch number will be sent to an outside testing laboratory. The above tests will be performed per the references in ASTM E 1417. The certification report, indicating acceptable results, will be on file in the NDT office for at least 10 years. This report shall be signed/stamped by a responsible BARI representative indicating acceptance of the results. (*Note: Item 22.1 relates to the standard's calibration requirements.*)

23.0: Penetrant tank cleaning (annually)

23.1 At least once per year the penetrant tank shall be checked for cleanliness. If necessary, pump out the penetrant within four inches of the bottom. The bottom four inches of penetrant will be removed and discarded. The tank walls and bottom will be thoroughly cleaned with an acceptable cleaner and allowed to dry. The removed penetrant may be reused if all tests and process control checks are acceptable.

23.2 After the tank is refilled with either the old penetrant or a new batch, all in-house process control checks will be accomplished and deemed acceptable prior to inspecting production parts.

24.0: Dry developer check (annually)

24.1 Push the start pedal on the developer tank and verify that the developer cloud in the tank is sufficient to properly cover the parts. Ensure that the developer is dry and not caked. Replace if necessary. Under ultraviolet light, if 10 or more fluorescent specks are seen in a four-inch diameter, the developer is contaminated and shall be replaced.

25.0: Rinse-water pressure and temperature tests (each shift)

25.1 The rinse-water pressure and temperature shall be checked with the water flowing through the nozzle. The required temperature and pressure shall be obtained after the flow has stabilized. A maximum of 30 psi is allowed, and the temperature shall be between 50°F and 90°F.

26.0: Drying oven temperature (each shift)

26.1 Drying oven temperature shall be checked after the oven has been turned on and the temperature allowed to reach the set point with doors or curtains closed. The temperature shall be maintained between 110°F and 140°F.

27.0: Black light filters (each shift)

27.1 All black light filters shall be checked to verify they are not cracked, chipped, or otherwise damaged. The filters shall be checked to ensure they are not contaminated with dirt, developer, and so on. Damaged filters shall be replaced prior to use, and dirty filters will be cleaned.

28.0: Black light intensity (each shift)

28.1 Black lights shall be allowed to warm up for a minimum of 15 minutes prior to performing the intensity check.

28.2 In the inspection booth, use the calibrated spectroline radiometer (or an equivalent) and measure the maximum intensity 15 inches from the black light filter. Manipulate the light as required to obtain the maximum intensity. The minimum intensity required is 1200 $\mu W/cm^2$.

28.3 In the wash booth, the intensity must be a minimum of 100 $\mu W/cm^2$ at the surface of the part being washed.

28.4 The ultraviolet borescope shall be checked prior to use with a calibrated meter. The intensity shall be a minimum of 1000 $\mu W/cm^2$ at the distance that it is to be used. Record intensity and distance in a log kept with the borescope. (Note: Item 28.4 relates to the standard's calibration requirements.)

29.0: Ambient/bright white light check (each shift)

29.1 The maximum ambient white light allowed in the inspection booth is two foot-candles.

29.2 The test is performed using a calibrated spectroline white light meter or equivalent. Perform the test with the black light on, in its normal position, and with the curtains closed. Place the meter on the part or on the surface where the parts are normally inspected and note the reading on the meter.

29.3 In the wash booth, place the meter on the inspection surface. With the black light on and the curtains closed, note the meter reading. A maximum of 10 foot-candles is allowed.

29.4 The intensity of the white light used for evaluating indications must be a minimum of 100 foot-candles. Test the intensity at the surface being inspected.

30.0: Air pressure checks (each shift)

30.1 If used, the air pressure check at the developer removal station and pre-drying oven shall be performed by opening the air nozzle and reading the pressure indicated on the appropriate gages.

30.2 The maximum pressure for removing excess developer is 5 psi. The maximum pressure to remove excess water from the part prior to placing in the drying oven is 25 psi.

31.0: Air line cleanliness (daily)

31.1 If used, each air line connected to a handheld air nozzle shall be checked by blowing the properly regulated air onto a white filter-type paper, paper towel, or mirror. There shall be no evidence of contamination visible.

32.0: Gage, meter, and timer calibration

32.1 Gages, meters, timers, and so on, used in checking the penetrant system process controls and equipment shall be calibrated semiannually by an outside calibration source. The drying oven controller and temperature indicator shall be calibrated quarterly. The equipment will be calibrated to the manufacturer's specifications, and the certification will be retained in the NDT office.

32.2 All measuring devices shall be calibrated through the range being used to include the minimum and maximum limits.

ANNEX 1: NAS 412

1.1: Precleaning instructions

Typically, parts shall be cleaned with the aqueous degreaser. Other methods may be used subject to approval of the customer. Parts shall be cleaned by immersion or spraying, depending on part size. The detergent used for cleaning shall be Brulin 1990 GD or an approved equivalent.

The temperature shall be maintained between 130°F and 150°F, and the concentration of detergent to water shall be between 8% and 25% by volume.

Immerse the part in cleaning solution and allow agitating or spray solution on parts until all surfaces are visually clean. When parts appear to be clean, rinse them with water until the surfaces are water-break free.

All surfaces shall be free of any contaminant or condition that could retain penetrant or mask defects. Abrasive blasting shall be used to clean metals only if the surface is not cleaned by the process or if surface defects are not covered or contaminated by the abrasive material.

Paint shall be removed by chemical means and not by abrasive methods. All chemical solutions must be neutralized and flushed from part surfaces.

All parts to be inspected shall be clean and dry. Parts shall be dried for 30 minutes at 130°F–200°F to ensure that any liquid in a defect has evaporated. A visual inspection shall verify that parts are completely dry prior to inspection. In no case shall the temperature exceed customer requirements.

Soft metals or metals that have been previously machined, sandblasted, peened, deburred, and so on, shall be etched to remove smeared metal that could mask defects. Parts requiring pre-penetrant etch shall be sent to an approved outside vendor (refer to customer specifications for particular precleaning requirements).

After drying, parts shall be cooled to below 100°F before penetrant is applied.

Appendix E
Sample Nondestructive Testing Procedure

Purpose

1.0 Boogaloo's performance of nondestructive testing (NDT) will be accomplished in accordance with operating procedures, inspection instructions, or manufacturers' manuals, as required. Inspectors must be properly certified for the work they are performing in accordance with an acceptable industry standard equivalent to the American Society for Nondestructive Testing's Recommended Practice No. SNT-TC1A, as well as meeting or exceeding the requirements for qualification and certification of NDT personnel as outlined in MIL-STD-41OE and NAS 410.

1.1 Equipment used for NDT inspections must conform to Boogaloo's operating procedures and inspection requirements, and/or those of the manufacturer, as applicable. No equipment substitutions may be made without the concurrence of the aircraft/engine or an appropriately rated individual or agency.

1.2 Reference standards must be certified to meet the specifications outlined by the manufacturer's manual. Certification must be made by the contractor's quality assurance department or the manufacturer of the standard.

1.3 The NDT of items can be carried out after a technician who is qualified in accordance with OP 0117.1 has satisfied the NDT requirements.

1.4 Limitations

1.4.1 The NDT eligible under this provision is restricted to liquid penetrant and magnetic particle methods that do not require interpretative skills beyond the training received. This introduction into your procedure sets the tone that you intend to comply with section 7.5.2 in all respects. This refers to applicable industry standards such as MIL-STD-41OE and NAS 410.

The training requirements should also be established and set forth. Do not rely solely on the terms of the industry standards themselves for the training requirements; they should be incorporated into your training matrix under section 6.2.2. An example of the qualification procedure is shown as section six.

SECTION SIX: PROCEDURE

6.0 Boogaloo NDT personnel shall be trained, examined, qualified, and certified in accordance with the requirements of this procedure before providing NDT services on behalf of this organization.

6.1 Boogaloo NDT personnel examinations and records of qualification and certification shall be complete and available for inspection by representatives of both prime contractors and appropriate government agencies.

6.2 At the option of Boogaloo, an outside agency may be engaged to provide NDT level III services. In such instances, the responsibility of certification is retained by Boogaloo. Boogaloo appoints the consulting level III as the "responsible level III," per NAS 410.

NDT METHODS

Qualification and certification of NDT personnel in accordance with this procedure are applicable to each of the following test methods:

- Liquid penetrant testing (PT)
- Magnetic particle testing (MT)

6.4.1: Levels of qualification and certification

6.4.2 The levels of qualification established by this procedure are as follows:

—Trainee

—Level I

—Level II

—Instructor

—Level III

—Auditor

The levels of certification established by this procedure are as follows:

—Level I

—Level II

—Level III

6.4.3 A trainee is an individual who is participating in a training program for an NDT method and is not certified. Trainees shall obtain work experience only under the direct supervision of a level II, a level III, or an instructor in the same method. Trainees shall not independently conduct tests, make accept/reject determinations, or perform any other NDT functions.

6.4.4 Level I is the first certifiable qualification level. The level I certification shall be for a specific technique in a given method. The level I individual shall have the skills and knowledge to perform specific tests, specific calibrations, and, with prior written approval of the appropriate level III, specific interpretations and evaluations for acceptance or rejection, and to document the results in accordance with specific procedures. The individual shall be knowledgeable of any necessary preparation of parts before or after inspection. The individual shall be able to follow procedures in the techniques for which certified and shall receive the necessary guidance or supervision from a certified level II or level III.

6.4.5 A level II shall have the skills and knowledge to set up and calibrate equipment and conduct tests, and to interpret, evaluate, and document results in accordance with procedures approved by the appropriate level III. The level II shall be thoroughly familiar with the scope and limitations of the method in which certified and shall be capable of directing the work of trainees and level I personnel. The level II shall be able to organize and document NDT results. The level II shall be familiar with the codes, standards, and other contractual documents that control the NDT methods in use at Boogaloo.

6.4.6 Instructors shall have the skills and knowledge to plan, organize, and present classroom, laboratory, or on-the-job training programs of instruction, in accordance with approved course outlines and in the method for which appointed. The individual shall be familiar with the codes, standards, and other contractual documents that control the NDT methods in use at Boogaloo.

6.4.7 Level III individuals shall have the skills and knowledge to interpret codes, standards, and other contractual documents that control the NDT methods in use at Boogaloo. In addition, level III personnel shall be capable of selecting the method and technique for a specific inspection, preparing procedures and verifying their adequacy, and conducting training and examinations of personnel in the method(s) certified. Only personnel certified to level III in the applicable method shall have the authority to approve procedures for technical adequacy. Level III personnel shall not conduct NDT for the acceptance of parts unless a demonstration of proficiency in this capability is included as part of the level III individual's practical examination.

6.4.8 Boogaloo may use an outside agency to develop a certification program, train and examine NDT personnel, and perform any other level III functions. An outside agency cannot certify personnel. Boogaloo shall document the suitability of the outside agency selected to perform any function to meet the requirements of this procedure. This documentation shall be sufficient to justify that the agency is capable of performing the required level III functions.

6.4.9 Auditors shall have the education, training, skills, and knowledge to understand the processes and procedures used in the application of NDT processes. They shall be familiar with the applicable codes, standards, and other contractual documents that control Boogaloo's

NDT processes. Another part of your procedure should show what examinations, whether practical, written, or physical, are required, as shown in the following section.

6.5.9: Examinations

6.5.9.0 The examinations to verify the physical and technical qualifications of candidates for certification shall consist of a physical examination, a general examination, a specific examination, and a practical examination. The questions employed for the general and specific examinations and the checklist for the practical examination shall be on file at Boogaloo and shall be made available to appropriate customer representatives upon request. All examinations shall be changed periodically so that no person takes the same exam more than once.

6.5.9.0.1 Written examination questions and test pieces used for practical examinations shall be changed periodically to preclude compromising the validity of the examination and to ensure that no test candidate ever receives the same exam.

6.5.9.0.2 The physical examination shall ensure that the candidate's near vision and color perception meet the following requirements. The near vision and color perception tests shall be administered annually by an individual approved by the Boogaloo level III. The level III's signature on the eye examination report signifies approval of the examiner and of the examination results.

Another area that should be addressed is the loss of certification and reinstatement procedures, as is illustrated in sections 6.6.2 and 6.6.3.

6.6.2 Certification may expire, be suspended, or be revoked. Certification shall expire when employment is terminated or when the certification interval has lapsed with no recertification attempted. Certification shall be suspended when the periodic physical examination is overdue, when the individual does not perform in the method certified for 12 consecutive months, or when the individual's performance is found to be deficient in any manner. Certification shall be revoked when the individual does not perform in the method certified for 24 consecutive months, when the individual is determined to be incompetent, or when the individual's conduct is determined to be unethical. It shall be the responsibility of the inspector to document his or her performance of inspection for the fields certified.

6.6.3 Certifications that have been suspended may be reinstated when the cause for suspension has been corrected and the correction has been verified by Boogaloo. Certifications that have expired or been revoked may not be reinstated except by recertification.

6.6.4 Level I and II personnel shall be recertified at intervals not to exceed three years. Level III personnel shall be recertified at intervals not to exceed five years. The physical and practical examinations equivalent to those required for initial certification shall be given prior to recertification. The extent to which the individual's knowledge of the general and specific examination subject areas is examined shall be determined by the Boogaloo level III.

It's a good idea to write the NDT procedure to include the requirements of the standard being used. If the organization is using the magnetic particle inspection (MPI) method of NDT, the procedure should include the purpose and scope as outlined in the following sections.

1.0: Purpose

1.1 The purpose of this procedure is to define the minimum requirements for the control and performance of MPI performed at Boogaloo.

2.0: Scope

This document applies to all NDT personnel qualified and certified to perform MPI per OP 0017.1. This procedure is mandatory and applies to all products inspected by the MPI method, including reworked parts. This document meets or exceeds the requirements of MIL-STD-1949, ASTM E 1444, and all applicable customer specifications. In the event of a conflict between this document and customer requirements, the customer requirements shall have precedence.

Customer products having specialized requirements will be detailed in supplements to this procedure. The applicable appendix shall be referred to for any processing or inspection parameters that differ from this procedure. In such cases, the customer requirements spelled out in the applicable appendix will have precedence.

Finally, make sure that you have devised the periodic testing required and control the forms for which you will use as records and verification of completion of these processes.

Appendix F
Sample Aircraft Turtle Diagram

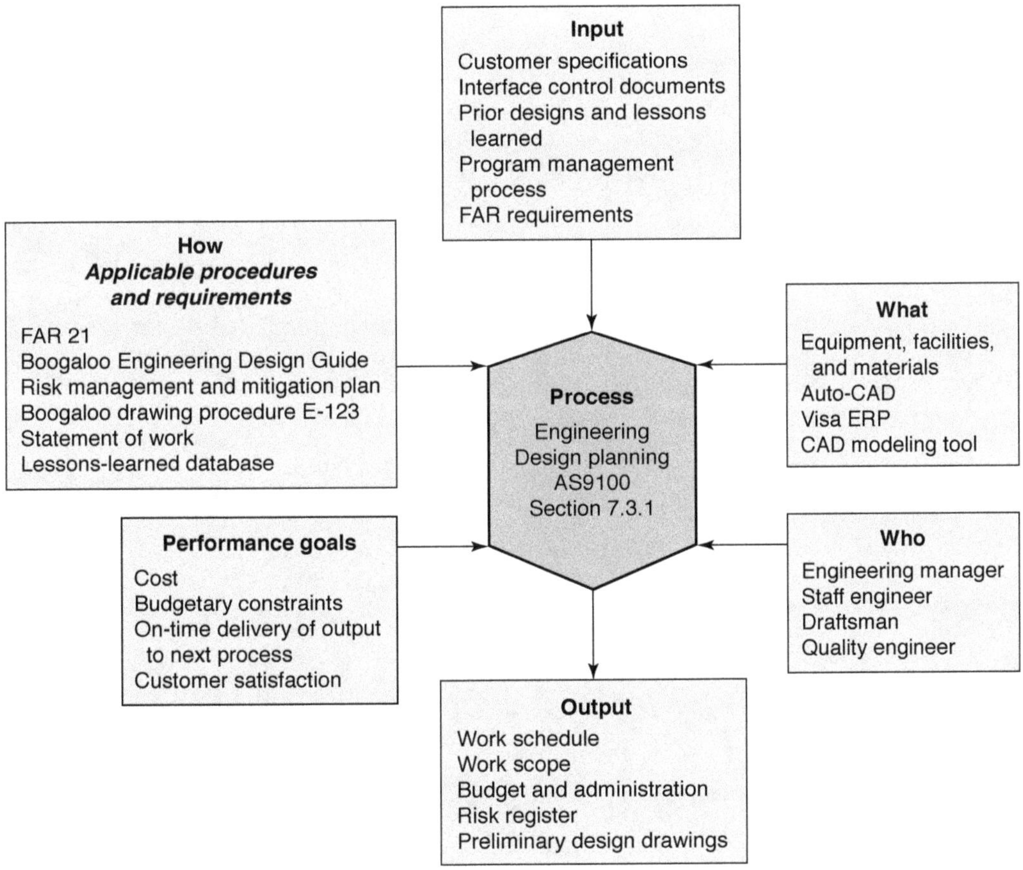

Appendix G
Summary of Changes to AS9100C from AS9100B

SECTION 4

AS9100C inserts some sentences and clarifies some language in section 4. The update adds customer and statutory/regulatory requirements to the aerospace quality management system (AQMS). In section 4.1, subclause (a), the word "identify" has been replaced with "determine."

The standard now expects the organization to do more than just "identify processes needed"; it expects that the organization will determine, after consideration, what processes are needed. The effect on section 4.1 is not substantial, and most of the language is to put the section in alignment with other changes elsewhere in the revised standard. For example, paragraph (e) is amended to show measurement, where applicable, by the organization.

This modification continues the reasoning that processes are to be not only identified but also defined and applied. Note 3 (also added with revision C) addresses this subject. The one existing note in revision B under clause 4.1 has been further clarified, and two new notes have been added.

Revision C adds note 2 to explain what an outsourced process is. Note 3 was added to ensure that outsourced processes are handled in accordance with section 7.4, which pertains to purchasing. Auditors have always considered outsourced processes to be requirements under section 7.4, so there should be no real difference between the revision B applications of the revision C changes.

Section 4.2.1: General

The changes in this section amount to inclusion of the word "records" of subclauses (c) and (d). In addition, subclause (e) was eliminated because the reference to records was inserted into subclauses (c) and (d).

AS9100C reworded the sentence after paragraph (f) and deleted the second sentence, as it makes sense that personnel not only have access but also should be made aware of documentation and changes.

Most of the changes in section 4.2.1 are administrative. A major change is the clarification that any requirement for a documented procedure be covered by more than one document if necessary; this allows a more common-sense approach to some of the requirements of the standard. For example, section 4.2.4 had and still has a requirement that the procedure contain the method for controlling the records created and/or retained by suppliers. It makes more sense that supplier requirements be handled in the documentation requirements listed in section 7.4.2 or 7.4.1 and not limited to section 4.2.4.

Section 4.2.2: Quality Manual

Revision C eliminated the aerospace requirement that followed paragraph (b). The International Aerospace Quality Group (IAQG) believed that the deleted sentence was too prescriptive and that its requirements were redundant.

Section 4.2.3: Control of documents

The only change to section 4.2.3 occurs in subclause (f):

> f) To ensure that documents of external origin are identified and their distribution controlled, and . . .

The change clarifies which external documents are necessary to be controlled:

> f) To ensure that documents of external origin determined by the organization to be necessary for the planning and operation of the quality management system are identified and their distribution controlled.

Only external documents necessary for the planning and the operation of the quality management system (QMS) are required to be controlled. The developers of AS9100C deleted the final sentence in this section because the rework to section 4.1 addresses the general requirements to include customers and statutory/regulatory authorities.

The change to subclause (f) will make it much easier for organizations to decide which external documents they need to control. If the document is used for planning or operations, such as process specifications, it will continue to be controlled. Other documents of external origin don't need to be controlled.

Section 4.2.4: Control of records

The opening sentence of clause 4.2.4 clarifies that records are not just to be "maintained" but controlled. The IAQG believed that simply maintaining records was too simplistic and that the records (and documents) needed to be controlled.

Other than the added requirement that records be controlled and the elimination of the word "time," the substance of section 4.2.4 remains the same. No changes were made to its content or application. The last sentence in this section was also eliminated because of the addition to clause 4.1 of the same requirements.

SECTION 5

Section 5.1: Management commitment

No changes.

Section 5.2: Customer focus

Revision C has added a requirement to measure product conformity and on-time delivery performance and take action if results are not achieved as planned. This requirement will affect most organizations. It will be an extension of the management review, and it will require action plans or similar tools to be implemented when on-time performance falls below the organization's goal. This will be true even if on-time delivery or product conformity aren't stated quality objectives.

Section 5.3: Quality policy

No changes.

Section 5.4: Planning

No changes.

Section 5.5.1: Responsibility and authority

No changes.

Section 5.5.2: Management representative

This subclause has been changed to align with ISO 9001:2008. It now requires that the management representative be a member of the organization's management. Subclause (d) was revised to add unrestricted access to top management and clarify the contact to pertain to quality management issues.

It's rare to see a management representative who isn't a member of his or her organization's management team. The change clarifies the requirement because there have been instances where the management representative is outsourced to a third party, perhaps a consultant. The change to paragraph (d) ensures that there is a direct line from the management representative to top management on quality management issues. This will prevent issues of top management allowing the shunting of contact to an intermediary person or group.

Section 5.5.3: Internal communication

No changes.

Section 5.6: Management review

No changes.

SECTION 6

Section 6.2.1: General

This clause has been amended to change the wording from work affecting "product quality" to work affecting "conformity to product requirements."

A note was added to explain the effect of the word "conformity."

Section 6.2.2: Competence, training, and awareness

The title of this clause has been changed from "Competence, Awareness, and Training" to "Competence, Training, and Awareness." The change in section 6.2.1 from "product quality" to "conformity to product requirements" has been aligned with paragraph (a) of this section.

Paragraph (b) was amended to include the phrase "where applicable" and clarify the existing phrase "satisfy these needs."

This section was changed to reflect that additional training may not be required if personnel are already competent in processes. Auditors have loosely applied this configuration in the past. A newly hired certified welder doesn't need training just for the sake of the training requirements imposed by the standard. This is another common-sense change. The phrase "satisfy these needs" that was included in the previous version of the standard was unclear and needed clarification. Generally, however, no significant changes were made in section 6.2.2.

Section 6.3: Infrastructure

The only change in section 6.3 was to add "information systems" to section (c), a change prompted by ISO 9001:2008. IT's placement in this section indicates that the electronic age has hit the AQMS. This section will require the implementation of IT disaster recovery plans, backup plans, and the like for the auditor to review. The rationale of the change is the insurance of saving the records and programs required by clause 4.2.4. This will take additional effort, but most companies already back up their data and have recovery plans.

Section 6.4: Work environment

The note at the bottom of this section was replaced with a note taken directly from ISO 9001:2008: "The term 'work environment' relates to those conditions under which work is performed including physical, environmental, and other factors (such as noise, temperature, humidity, lighting, or weather)."

The change helps align AS9100's requirements with those of ISO 9001. They will have no significant effect on the revision C process.

SECTION 7

Section 7.1: Planning of product realization

The 2008 revision to ISO 9001 added "measurement" as one of the required activities to be determined during the planning of product realization. One of the most significant changes in revision C is the moving of configuration management from section 4.3 to section 7.1.3. Auditors will assess configuration management not as a stand-alone process but as part of product realization. This is a better methodology for handling this critical aerospace requirement. Refer to the revision B guidance on configuration management in section 4.3 for a discussion of the elements involved. Section 7.1.3 represents the real change; the effect on section 7.1 sets up the use of configuration management as a tool.

Section 7.1.1: Project management

Revision C has added three new subclauses to be incorporated into the planning phase of the product (remember that section 7.1.3 was clause 4.3 in the previous edition).

This is a lead-in paragraph for the really big change on the increased use of risk analysis as part of the AQMS, which is discussed in section 7.1.2.

Section 7.1.2: Risk management

In AS9100B, "risk" was only preliminarily used in clause 7.2.2 (d) to review product requirements for risks such as new technology and short delivery times. Revision C has increased the focus on the risks involved during product realization planning. It will mean more effort.

One of the additions in revision C concerns section 7.1.2, Risk management. In any business, a risk analysis should be performed as part of the overall process for each contract. The resulting data will identify potential issues and risks before they can pose cost and/or scheduling problems. Mathematically, this analysis equates to risk = probability of event multiplied by the cost of the event.

NASA work instruction 170846 contains an excellent description of risk analysis.

The risk analysis can be depicted on a simple matrix, such as the one shown in Figure 6.2, which describes the order of probability of a risk event against the cost of the event.

Numerous publications describe risk assessment and demonstrate how to calculate it. A large number of these documents are in the federal government's space program (Department of Defense and NASA) and its prime contractors (Jet Propulsion Laboratory, Lockheed Martin). The management program discussed in section 7.1.3 is not as detailed as a space vehicle launch; the risk most organizations face is their ability to deliver their products or services and what factors may impede that delivery. The examples supplied in revision B are straightforward: "New technology, short delivery time scale." Revision C includes an outline of what a risk management program should contain:

- Assignment of responsibilities for risk management
- Definition of risk criteria (e.g., likelihood, consequences, risk acceptance)
- Identification, assessment, and communication of risks throughout product realization
- Identification, implementation, and management of actions to mitigate risks that exceed defined risk acceptance criteria
- Acceptance of remaining risks after implementing risk mitigating actions

It's important to note that section 7.1.2 is a scheme requirement, and each of the program requirements listed in paragraphs (a)–(e) will be verified at audit. Appendix B contains a complex risk management program. Implementation must include the necessary forms, templates, reports, record keeping, and training.

Section 7.1.3: Configuration management

Revision C has moved this clause from section 4.3 to 7.1.3 and redrafted the requirement to include aspects of a configuration management program.

Other than its relocation and the addition of elements of a configuration management program, little has changed regarding this requirement from revision B to revision C. The discussion in revision B should give adequate guidance on the revision C changes.

Section 7.1.4: Control of work transfers

Revision C has relocated the requirement for the control of work transfers from clause 7.5.1.4 to 7.1.4.

The only significant change clarifies the types of work transfers and notes that it could be a temporary or permanent transfer.

Section 7.2.1: Determination of requirements related to the product

In section 7.2.1, the word "related" was eliminated and the word "applicable" was added.

The word "determined" was removed in paragraph (d) and "considered necessary" was added.

Two notes were also added concerning special requirements and post-delivery activities.

ISO 9001:2008 makes post-delivery customer-satisfaction activities a requirement. The writers of the standard removed the word "determined" because it already appeared in the lead-in sentence and was considered redundant. The "considered necessary" addition was also made in ISO 9001:2008.

The first note added was a result of the 2008 revision of ISO 9001. The effect of the changes to section 7.2.1 of the AQMS is minimal. Post-delivery activities, however, will be the subject of audits conducted under revision C.

Revision C has also added the second note as an aerospace requirement. The reference to "special requirements" can be found in section 3.2 of revision C.

Section 7.2.2: Review of requirements related to the product

This section was amended by moving paragraph (d) to paragraph (e) and linking the relocated paragraph to the new section 7.1.2. Paragraph (d) now refers to the clause in section 3.2 that defines special requirements.

There should be minimal effect on this section, other than a close adherence to section 3.2, which addresses special requirements.

This section is closely aligned with the risk management requirements in section 7.1.2.

Section 7.2.3: Customer communication

No changes.

Section 7.3.1: Design and development planning

A significant change to section 7.3.1 involves the relief granted by the new note in section 7.3.1 (b). Instead of requiring that design and development review, verification, and validation have separate records, the new requirement allows the organization to maintain records separately or combine them in a way that's suitable for the product and the organization. This allows smaller organizations to conduct validation and verification concurrently. Records must still be maintained, but the organization can streamline implementation of the planning process for design and development. The rest of the changes are administrative in nature and have no real effect on design and development except to add in regulatory considerations.

Section 7.3.2: Design and development inputs

The change to this clause is simply the removal of the word "these" and replacing it with the word "the."

There is no effect on the changes to this section.

7.3.3 Design and development outputs

For the most part, the changes to this section are administrative. They tie the new definitions in section 3, including "critical items," to the output requirements. The effect on this section merely extends the requirement for identification of key characteristics to include critical items. This may result in additional procedural requirements in the organization's design guide and an additional requirement for audits.

Section 3.3: Critical items

Those items (e.g., functions, parts, software, characteristics, processes) having significant effect on the product realization and use of the product; including safety, performance, form, fit, function, reproducibility, service life, etc.; that require specific actions to ensure they are adequately managed. Examples of critical items include safety critical items, fracture critical items, mission critical items, key characteristics, etc.

Section 7.3.4: Design and development review

No changes.

Section 7.3.5: Design and development verification

The note from revision B was deleted in revision C.

Section 7.3.6: Design and development validation

The note in revision B was deleted in revision C.

Sections 7.3.6.1 and 7.3.6.2: Design and development verification, validation testing, and documentation

The writers of revision C joined section 7.3.6.2 from revision B with section 7.3.6.1.

Discussion of 7.3.5–7.3.6.1 Revision C Changes

The changes in these sections are designed to join verification and validation as continuous processes. In practice, there are no substantial differences in the application of the two base clauses, 7.3.5 and 7.3.6.

Section 7.3.7: Control of design and development changes

The writers of ISO 9001:2008 revised the first and last paragraphs of this section into a single paragraph without making editorial changes.

The changes in section 7.3.7 were implemented because of the increased focus on the configuration management program. Clause 4.1 now requires that applicable statutory and regulatory requirements be addressed throughout the QMS, so the deleted sentence became redundant. The only change in this section is designed to ensure that design and development changes are controlled by the configuration management process. As most organizations already adhere to this rule, there is little meaningful change in this section.

Section 7.4.1: Purchasing process

The changes to this section clarify the information to be included in the approved supplier list (ASL) and require that the *results* of performance reviews, rather than the records of the reviews, be used for controlling suppliers. These changes are rather insignificant and are designed to help auditors and organizations compile their ASL. The change to paragraph (e) is an increase in the requirements for supplier control procedures. It also strongly implies that this is a gray area for vendors that may be used on a limited basis (controlled use), consistent with their level of approval status.

Paragraph (f) was changed to continue the focus of risk management newly defined in section 7.1.2. The changes are an attempt at clarifying the control of suppliers, but little additional work specific to section 7.4.1 will be necessary.

Section 7.4.2: Purchasing information

The changes in section 7.4.2 are mainly editorial. The new paragraph (h) ties in to section 4.2.4's requirement of record retention for suppliers: "The documented procedure shall define the method for controlling records that are created by and/or retained by suppliers."

This has also been addressed to allow the procedure to reside in several procedures, one of which is the purchasing procedure and is consistent with the new relaxation provision in section 4.2.1, which says: "A single document may include the requirements for one or more procedures. A requirement for a documented procedure may be covered by more than one document."

This is a good common-sense change. The other changes are consistent with the standard's universal alignment of the term "supply chain" rather than "suppliers." The change to appease some supply chain members about the right of access was clarified to require limited, rather than full, access. The addition of "critical items" is in alignment with section 3.3's new requirement:

> Those items (e.g., functions, parts, software, characteristics, processes) having significant effect on the product realization and use of the product; including safety, performance, form, fit, function, reproducibility, service, life, etc.; that require specific actions to ensure they are adequately managed. Examples of critical items include safety critical items, fracture critical items, mission critical items, key characteristics, etc.

The changes should not reflect any additional effect on the procedures and practices of the organization and are much more clear than the requirements in revision B.

Section 7.4.3: Verification of purchased product

This is a good relaxation of some revision B requirements, especially the elimination of the periodic validation requirements. The rest of the changes are for syntax and editorial purposes. This section should require less effort in the long run.

Section 7.5.1

Revision C reconfigured the opening requirements by moving the planning section to the bottom of the clause and making some editorial changes.

Paragraph (d) was modified to align with the changes to ISO 9001:2008, which refers to "monitoring and measuring devices" instead of "monitoring and measuring equipment." This change is also reflected in section 7.6.

The spirit and intent of revision C's section 7.5.1 remains as it was in revision B. Most of the changes were made to clarify language and align AS9100 with ISO 9001:2008.

Section 7.5.1.1

Revision C has eliminated the revision B paragraph described above for production documentation. It has been replaced with section 7.5.1.1 and is titled "Production Process Verification." The old requirement (a) has become a note under section 7.5.1 (a), and the old requirement (b) has become a new note under section 7.5.1. The requirements of revision B's section 8.2.4.2, First-article inspection, have been moved to revision C's section 7.5.1.1.

There has always been some question as to whether the first-article inspection process belonged in section 8. It's not a monitoring and measuring process, but a process used to ensure that a product is manufactured under controlled conditions. Further, because revision B had the first-article inspection requirement in section 8, the organization couldn't properly exclude the requirements of section 8.2.4.2. (Remember that only clauses in section 7 are subject to exclusion.)

This became a minor arguing point between auditors and auditee organizations. The placement of the first-article inspection requirements under section 7.5.1.1 corrects any misconceptions about the excludability of the first-article inspection requirements. In practice, this change affects only the excludability of the first-article requirements, not the subject matter of first-article inspections.

Section 7.5.1.2

The biggest change to this section is found in the new standard AS9101D. Section 4.2.2.8 of AS9101D requires that when special processes are identified in the audit plan, the audit team "shall evaluate process validation, as well as the monitoring, measuring, and the control of these processes." Several other requirements in section 4.2.2.8 pertain directly to the special processes area.

Revision C made editorial changes to the contents of section 7.5.1.2 by eliminating its second paragraph. This change continues into the third paragraph, where the standard refers to "software programs" rather than "software." The revision also references "product conformity" rather than "product quality."

What might appear to be a major change in this section really isn't. Although the language of section 7.5.1.2 seems to have eliminated the requirement for procedures to control implementation of process changes, it has not. Organizations are still required to control changes affecting processes, production equipment, tools, and software programs. This change was made because it likely appeared redundant with the requirement that a procedure be available specifically to control implementation. In practice, there is little difference between the revision B and the revision C requirements for the control of production changes. Taking that into consideration, the best practices listed for this section in the revision B standard still apply.

Section 7.5.1.3

The title of this section was amended in revision C to reflect "software" programs rather than the more specific "numerical control" programs. Other changes include the elimination of the direct reference to first-article inspection and the substitution of the word "established" for the word "defined" in the last sentence.

The changes to this section, although they appear extensive, are relatively minor and clear up some conflicting and confusing language. The prescriptive requirement that validation include a first-article inspection has been eliminated as a specific requirement. This does not, however, eliminate the requirement of first-article inspection if that is the only method for verifying the validity of the product and for validating equipment and software prior to production. In practice, not much has changed with regard to the control of production equipment tools and software programs.

Section 7.5.1.4

Revision C has eliminated section 7.5.1.4 as a stand-alone clause to AS9100. The requirements contained within the section have been moved to clause 7.1.4. In its place, the updated standard includes section 7.5.1.4, Post-delivery support. This change is in keeping with changes to ISO 9001:2008, which requires post-delivery support activity. Also missing from revision C is section 7.5.1.5, Control of service operations.

The elements of the revision B requirement for control of service operations are now contained in this section, with some changes.

The requirement for control of work temporarily transferred outside the organization's facilities has been moved to section 7.1.4 and still applies. In its place, a requirement for post-delivery support has been implemented. This replaces the section on control of service operations. In too many previous instances, there was a significant difference of opinion as to what constituted a "service."

ISO 9001:2008 gives increased focus to post-delivery support; the old section 7.5.1.5's requirements are almost completely replicated in the new 7.5.1.4 section. There are only minor differences. This will clear up the question of what constitutes "service" and will eliminate the ongoing differences of opinion on this subject between registrars and organizations.

Section 7.5.2

Section 7.5.2 didn't receive any significant changes in revision C. It clarifies that a process that produces products that have deficiencies that can be identified only after the product is in use or the service has been delivered is a special process.

There was also a small change to the note: "These processes are frequently referred to as special processes."

Finally, the aerospace-specific additions to paragraphs (a) and (c) were removed.

For the most part, the revisions to section 7.5.2 are minor and include no additional requirements. However, there are significant changes in the application of AS9101D in the area of special processes, which are addressed in its section 4.2.2.8. This applies more to registrars and auditors than to organizations.

Section 7.5.3

Organizations must identify their products by suitable means "throughout product realization." The requirement for a documented process for acceptance authority has been changed to "appropriate" controls and the records have been amplified.

This requirement has been changed to a note ("shall provide" to "can include"), as the text is prescriptive and not applicable to all organizations. The revision removes the contract and regulatory references. This is consistent with the overall strategy to state the requirement in section 4.1 and have it reflect incorporation throughout the standard.

The most significant change to this section is the elimination of the requirement for a "documented" control for stamps, signatures, passwords, and so on. The revised standard requires "appropriate" controls as an acknowledgment that there are many ways in which media can be controlled and that they should be "appropriate" to the use, custom, and requirements of the organization.

Despite the standard's elimination of the requirement for a documented control for stamps, signatures, passwords, and so on, it's still a good practice to have a process that takes into account the use of stamps, lost stamps, stamp audit programs, and the return of stamps to reduce variation (and error) in daily operations. This is especially true if the operation involves flight operations or has FAA oversight.

Section 7.5.4

A grammatical change was made to section 7.5.4 that has little effect on the organization's handling of customer property.

The note has been modified and is in alignment with ISO 9001:2008. The overall clause note has been simplified.

The changes to this section are generally minor. There is no change in the organization's responsibilities and duties with respect to section 7.5.4.

Section 7.5.5

The introduction to section 7.5.5 has been amended to ensure that conformity to requirements is included in the definition of what preservation of the product is all about.

To be consistent with ISO 9001:2008, this section was amended to add the word "requirements," and the requirement for accompanying documentation was deleted from this section and moved to section 8.2.4.

Most of these changes are administrative in nature. The deletion of the last paragraph in section 7.5.5 has no effect, because the requirement was moved to section 8.2.4.

Section 7.6

The title of this section has been changed from "Control of Monitoring and Measuring Devices" to "Control of Monitoring and Measuring Equipment." The term "equipment" was already used in several places in section 7.6, and ISO 9001:2008 made the change to the title as well.

The first aerospace requirement was amended to refer to "equipment" instead of "devices" and to include the word "verification."

A minor grammatical change was also made to the second aerospace-specific clause:

> The organization shall ensure that environmental conditions are suitable for the calibrations, inspections, measurements, and tests being carried out.

The words "or both" were added to section 7.6 (a) to ensure that monitoring and measurement equipment that can only be verified—not calibrated—is included (e.g., a go/no-go gage).

A minor change was made to paragraph (c) to reflect that identification of equipment might already be performed.

The recall requirement from paragraph (f) was reworded and emphasized in revision C.

A note was added to guide organizations on software confirmation, and a reference to outdated standards was removed.

Although the changes to this section appear to be extensive, this is not the case. Most of the changes were designed to reflect the changes in ISO 9001:2008. The aerospace-specific clauses have been amended to allow tools—such as go/no-go gages, for which only verification is available—to be comfortably included in the section on control of monitoring and measuring equipment.

Finally, software confirmation has been given some guidelines with the last note in revision C. If anything, this section's requirements are better defined and more clearly stated.

SECTION 8

Section 8.1

The word "requirements" was added in paragraph (a).

The note in revision C's version of section 8.1 was edited to change "may" to "can," four of the entries under "process control" were indented, and "criticality" was added to the last paragraph. The section entitled "inspection" was revised to eliminate what the IAQG recognized as the prescriptive verbiage.

The changes to this section are administrative. The addition of "criticality" merely changes the analysis from failure mode and effects analysis (FMEA) to failure mode, effect, and criticality analysis (FMECA). The changes don't affect the original spirit and intent of this section.

Section 8.2.1

Section 8.2.1 received significant additions because it defines mandatory measures that the organization must monitor. A new requirement and note have been added.

Organizations' customer satisfaction and management review process must be expanded to include all of revision C's new requirements, not just on-time delivery and customer complaints. Most, if not all, of these metrics can be obtained through internal sources, but, as the note points out, customer surveys are a great source. These items must be discussed at the management review meetings, and related charts and data should be presented at the meeting as objective evidence of the organization's progress with customer satisfaction.

Section 8.2.2

Subclause (a) in revision C contains a new note that clarifies that "planned arrangements" includes customer contractual requirements.

Minor text changes were made to the language of the section, and the entire section has been shortened. A change to the following paragraph clarifies that actions intended by the clause are "any necessary correction and corrective" actions.

The requirement for detailed tools was eliminated because it was viewed as too prescriptive. The note directing readers to ISO 10011:1, ISO 10011:2, ISO 10011:3, and ISO 19011 for guidance was updated to reflect current standards.

In practice, very little has changed with regard to the application of section 8.2.2. The most significant change involves the elimination of the requirement for detailed tools and techniques. It was decided that the requirement was too prescriptive and that the note to seek guidance from ISO 19011 was sufficient. It's still highly recommended that the organization develop some form of standardized tools and techniques to reduce variation within the internal audit program. The revisions to section 8.2.2 shouldn't create additional work.

Section 8.2.3

The phrase "to ensure conformity of the product" was removed from section 8.2.3 because not all processes directly affect product conformity. Thus, the requirement to monitor and measure processes should be expanded to apply to all processes, whether they affect conformity or not.

A note was added to provide guidance on the appropriate processes that may be involved in monitoring and measuring processes.

In the event of a process nonconformity, paragraph (c) was added and paragraph (d) was modified to include the word "any."

The changes to section 8.2.3 may require additional work from compliant organizations. In the past, monitoring and measuring tasks were limited to specific nonconformities. The changes in revision C make it clear that organizations must determine whether a process nonconformity is limited to a specific case or whether it could have affected other processes or products. This will cause the auditor to look at the analysis and results of process nonconformities. An additional requirement to determine a process nonconformity should be included in the organization's corrective action forms and process.

Section 8.2.4

The first paragraph in section 8.2.4 has been modified by moving the ISO 9001:2008 requirement concerning evidence of conformity to the bottom of the lead-in paragraph.

The section that addresses measurement requirements was imported from section 8.2.4.1 with no additional changes, other than alignment.

The section that addresses key characteristics was amended in revision C to include "critical items." There is also a new requirement that processes must be in place to control and monitor critical items and key characteristics that have been identified:

> When critical items, including key characteristics, have been identified, the organization shall ensure they are controlled and monitored in accordance with the established processes.

One of the most significant changes to section 8.2.4 concerns the use of sampling inspection or sampling plans. The revision B requirement that the sampling plan preclude acceptance of lots whose samples have known nonconformities was eliminated in revision C. In its place is a better description of what an appropriate sampling plan requires.

The sentence concerning product release and service delivery has been changed and clarified to include the customer.

The ISO 9001:2008 requirement for proper product delivery is reflected in revision C to show that only product "for delivery to the customer" requires an identified person to authorize product release.

Finally, the requirement that the organization ensure that all documents that are required to accompany the product are present at delivery has been moved to section 8.2.4 from section 7.5.5.

The changes in clarification of the various clauses and the consolidation of sections 8.2.4 and 8.2.4.1 simplify the application of product monitoring and measurement. No substantive changes were made to this section, with the exception of the clarification of the sampling plan and the addition of the "critical items" that would require additional focus and effort.

Section 8.2.4.1

Revision B's section 8.2.4.1 was eliminated and consolidated with section 8.2.4 in revision C. The section on test records was deleted.

The last paragraph of revision B's section 8.2.4.1 was moved to the next-to-the-last paragraph in revision C's section 8.2.4.

Discussion of Changes to Section 8.2.4.1

Although this section has been eliminated, it appears in section 8.2.4.

Section 8.2.4.2

This entire section has been relocated to section 7.5.1.1. In revising section 8.2.4.2 it was determined that first-article inspections are more of a production monitoring device, and this section was more appropriate for control of production.

Section 8.3

The changes to this section bring it into alignment with the changes to ISO 9001:2008's language (e.g., "can" instead of "may"). The clarification and streamlining of the requirements show that nonconformities that depart from customer requirements must be specifically approved by the customer, not just the organization. Another change is the addition of the containment requirements of paragraph (e). This will require changes in the documented procedure to ensure containment actions are appropriate to the effect of the nonconformity.

The organization's documented procedure for nonconforming product must include both provisions for customer approval in its matrix and planning for containment actions. Examples of containment actions include stopping a machine from operating when it's discovered that the machine is out of tolerance, and identifying a particular material as the cause of a nonconformity; supplier action is necessary.

Section 8.4

The only changes to this section include updating the reference specifications to the correct paragraph in the new revision. There are no changes with regard to substance and practice.

Section 8.5.1

This section has been improved to ensure that continual improvement projects and actions are monitored and evaluated for effectiveness. A new clause and a note have been added to this section.

This continues to be a very important clause in the overall scheme of both ISO 9001:2008 and the AQMS standards. Instead of just requiring general continual improvement of the effectiveness of the QMS, the additions require that specific actions be taken to monitor the implementation of improvement activities and to evaluate the effectiveness of those actions. The new note in section 8.5.1 includes examples where opportunities for improvement can be identified. This is a minimal list and should not be construed as a limit of improvement opportunities. (The best practices identified earlier in this section continue to be the best guidance available for continual improvement.)

Section 8.5.2

The language in section 8.5.2 and section 8.5.3 was changed as a result of changes in ISO 9001:2008. The changes include the words "cause" and "causes" to match with "nonconformities."

Likewise, the requirement that the documented procedure include a provision for "reviewing" was revised to show that the mere act of checking to see whether actions were taken was not the intent. Rather, the intent of this clause is to determine whether the action taken was effective.

The aerospace-specific paragraph (g) was clarified to show it wasn't just the "root cause" but also the "nonconformity."

Revision C has added paragraph (i) to require the organization to go beyond the detected problem and see whether more nonconforming product exists, and then take appropriate action.

The bulk of the changes in this section are minor. The change in paragraph (g) clarifies that the supplier is responsible for the root cause and nonconformance. The only true addition to the requirements of section 8.5.2 concerns the addition of paragraph (i) and the requirement that the procedure include a process for determining whether additional nonconforming product exists based on the causes of existing nonconformities and the taking of action when required on those additional nonconforming products. This will require additional effort on the part of the organization to implement such a process in its corrective action procedure. The organization's corrective action form should contain a link to its actions for determining whether additional nonconforming product exists, and include a discussion of action taken and completed.

Section 8.5.3

The requirement that the documented procedure include a provision for "reviewing" in paragraph (e) was clarified to show that the act of checking whether actions were taken was not the intent of the revision. Rather, the intent of this section is to determine the effectiveness of a subject to achieve established objectives.

Risk management has been added to this new note, along with several other tools typically used in preventive action processes. This list should not be construed as being limited to these few examples. An organization's risk management program, which is required by section 7.1.2, should include reference to section 8.5.3 in the area of identification, implementation, and management of actions to mitigate risks that exceed the risk acceptance criteria.

About the Author

James Culliton has worked in the aerospace and defense industries for 40 years. He has experience as a chief inspector and management representative for an AS9100/FAR 145–certified landing gear overhaul and repair facility. He was the director of quality and engineering in an aircraft manufacturing, certification, and modifications company and acted as an FAA-designated engineering representative. He has also worked as a maintenance station manager for a FAR 121–compliant airline. He holds current certification for AS9100C, AS9110A, and AS9120A and is listed on the OASIS database as a resource on those schemes. Culliton is the aviation, space, and defense associate program manager for a major registrar.

Index

Note: Page numbers followed by *f* refer to figures; those followed by *t* refer to tables.

A

B

C

D

NOTES

NOTES

www.ingramcontent.com/pod-product-compliance
Lightning Source LLC
LaVergne TN
LVHW062124200926
848484LV00007B/1759